Copyright and Redistribution Notice

This book, including all written content, images, and audio recordings, is protected under U.S. and international copyright laws. No part of this publication may be copied, reproduced, distributed, or transmitted in any form or by any means, including but not limited to electronic, mechanical, photocopying, recording, or other methods, without the prior written permission of the author or copyright holder, except in cases permitted by fair use or statutory exemptions.

Personal Use License

By purchasing this book, you are granted a **non-transferable, single-user license** to access and download the content for personal use only. You may not share, resell, or distribute this content in any form, whether for commercial or non-commercial purposes.

Digital Watermarking & Anti-Piracy Measures

This book may contain unique identifiers (such as your purchase details) embedded within the digital files. These measures help trace unauthorized distribution and protect the rights of the author.

Enforcement & Legal Action

Unauthorized redistribution of this content, whether in whole or in part, constitutes a violation of copyright law and may result in legal action, including but not limited to monetary damages, statutory penalties, and injunctive relief. The author actively monitors for copyright infringement and takes necessary steps to enforce their rights.

For licensing inquiries, bulk purchases, or permissions beyond the scope of this notice, please contact:

S. Clay Turner
https://sclayturner.com

"The thoughts and opinions expressed in this book are entirely my own. They do not necessarily reflect the views of Soaring Titan, Inc. This publication is made independently of Soaring Titan. References to Soaring Titan are made to either describe a historical fact, technical approach, or to convey my own interpretations and musings."

— *Clay Turner*

Table of Contents

Preface 1
Getting Oriented 4
    Understanding Intelligent Machines 4
        The Origins of Intelligent Machines 4
        The Current Status of Intelligent Machines 15
        The Future of Intelligent Machines 23
    Seeing Our Human Limitations 29
        Through the Camera Lens of a Machine 29
        Through Our Own Eyes 30
    Setting Our Expectations 31
        The Journey Ahead 31
        Achieving First Flight 32
The Three Stages of Superhuman Formation 35
    The Seeker 35
        Gottfried Wilhelm Leibniz (1646 — 1716) 35
        Experiencing the Spark 37
        Getting the Basics 38
        Full Formation 45
        Meet the Modern Titans 64
    The Maker 66
        Alan Turing (1912 — 1954) 66
        Working the Forge 69
        Getting the Basics 70
        Full Formation 88
        Meet the Modern Titans 98
    The Orchestrator 100
        Geoffrey Hinton (1947 — Present) 100
        Taking Flight 103
        Getting the Basics 105
        Full Formation 113
        Meet the Modern Titans 123
A Call to Reinvention 125
    A Social Imperative 125
        Evolving Social-Economic Structures 126
    A Personal Mandate 130
    Philosophy of the Soaring Titan 131
Additional Reading 134
Index 135

Superhuman: Mastering AI Agency to Exceed Personal Limits, Edition 1.0.0
© 2025 • S. Clay Turner • All Rights Reserved

Preface

In early July, 2023, I was standing on a beach in Sagres, Portugal, and staring back across the Atlantic towards home. My two, teenaged sons, Gabriel and Jude, were climbing the rockface above, glancing back occasionally for any hint of disapproval before climbing higher still. My wife, Carol, and daughter, Sarah, were playfully walking along the sand as the waves chased their feet. Yes, the family was "in the moment." I was on a business call, but I still couldn't help but contemplate the sheer distance between me and the voice on the other side as I gazed over the expanse of shifting water that scattered the sunlight so beautifully to the horizon. Technology is both wonderful, and wonderfully distracting, isn't it? This was, nonetheless, an important and fateful call — one where a key decision would be made of many crucial decisions to come. If I was to continue to pursue a technology career, then it would have to be exclusively in the realm of Artificial Intelligence, and more specifically, in its application. The decision had been made, then and there, with my friend and prior business partner, Michael Browning. Three years earlier, in 2020, Michael and I exited a SaaS company that we co-founded in 2012. Onovative, as it was called, had been a success — *generally* — and though nothing to retire on, the liquidity had both shielded my family from the effects of a global pandemic and bought an opportunity for a bit of "reinvention."

Prior to that call, I had been in Europe for nearly two months already. I'd arrived there in May with the kids to join Carol in the United Kingdom. After two prior stays and months apart, she was finally graduating in June with a culinary diploma from Le Cordon Bleu, London. We had lived on two separate continents during her studies and now it was time to reconnect and celebrate. Funny enough, we still had to live in two separate flats for the next month. Carol was living in Camden when we arrived, and we were now one tube stop away in Kentish Town. It was great, really. It was like dating Carol again — sneaking away to her place. The kids got a real kick out of this and would snicker at my late returns. The whole experience established new friendships and even expanded my own professional network nearly three fold. London has an extraordinarily accessible ecosystem. After we celebrated Carol's accomplishment, we left for Amsterdam, rented a car, and then drove along the northern coast for days before arriving in Normandy for a stay in the French countryside. We then drove back to Paris and flew to Seville, Spain. That night, we had tapas with Ivan and Alina — Ukrainian friends who had settled in Alicante with their young daughter just before Russia's cruel invasion. Ivan had established a software company there, and I was his first customer. We remained in Seville for several days before driving on to that beach in Sagres. After I had finished my call with Michael, my thoughts then turned to Lisbon; the last stop before walking 180 kilometers from Porto to Compestella, Spain, for El Camino de Santiago, the "Portuguese Way." Carol and Sarah flew home the same morning that Gabriel, Jude, and I took the train to Porto. On that long trek towards Santiago, I had plenty of time to reflect on what was behind me… And to what I knew would be a changed landscape in front of me upon return to the United States. I recognized, at least in part, that the shielding effect of that 2020 acquisition was nearing an end, but I had no idea just how challenging re-entry would be — even as (or especially as) an experienced technologist and entrepreneur.

Nearly four years earlier, when Michael and I entered into talks to sell our company, I was ready for change. Most founders love the excitement of creating a company, but things get complicated once you have investors, customers, employees, partners, KPIs, and… Yuck. I had already begun working on the next SaaS idea in anticipation of moving on from Onovative, even as the specter of a pandemic began to make its way across Europe, and into the US. If you're unfamiliar with the Dunning-Kruger Effect, then I suggest you take time to review it before "moving on to the next project." I had embarked on a new SaaS adventure with a dear friend and a beautiful soul, but one who lacked a technology background. We were going to conquer the world with our "unique blend of experience and talents." Within the first 6 months, it became clear to me that we had fallen from the "Peak of Mount Stupid" and into the "Pit of Despair." Building competencies on the long road back to having something viable would prove too difficult an endeavor. Time, money, and a friend were lost in the process, but I had explored angles to salvage the effort through a timesharing model with Ivan and his burgeoning team of software devs in the EU. This led to another SaaS pursuit with yet another friend, but this friend was (and is) a career technologist. This produced a better outcome, but Product Market Fit just wasn't coming easily there either. Now into 2023 with one SaaS brick, another unproven SaaS product nearly complete, and still a third SaaS product dragging along while SaaS valuations plummeted worldwide, I had resolved to make one final effort to turn things around. I would throw my hat into the Fractional CTO space for income, make final peace with SaaS, and pursue an AI venture with Michael called "Soaring Titan." I had even created my own philosophy around this idea of a "Soaring Titan."

It wasn't long after my return from Europe, however, that still more reality would set in. My "grand plan for recovery" was being challenged by the very thing that I'd been shielded from during the pandemic. Connecting with other "fractionals" quickly began to feel more like a flooded job market of highly skilled labor — *people like me* — all wanting the same thing. I had always created and sold "my inventions," but now I was having to sell "myself." Contracting for income proved futile. I had never had any issue with contracting in the past. Even applying for conventional roles were going nowhere. Although it offered little solace in the form of income, the drought provided just the kind of time required for soul-searching, networking, and R&D to prepare myself for integration into a society that would increasingly be filled with intelligent machines.

At time of writing, Salesforce has announced that it will no longer hire junior software developers, Microsoft has declared SaaS "dead," and 2025 is predicted to be the year that AI Agents enter the workforce. I concur, and I acknowledge that very few people have had the same opportunity that I've had to reinvent themselves or to prepare their children for this new reality. It is with this burden that I offer what I can in the pages that follow. I am not here to sell you AI hype. I'm not really promising riches if that's what you're here for. Let me be very clear from the beginning… To be superhuman is to rediscover what makes us uniquely human. This is, I believe, our best hope to survive and to thrive. This is what we will explore together in our own, brief pilgrimage. One of the biggest temptations with AI is to quickly grab an answer and to run with it, capturing only one small part of a bigger, more beautiful story. The whole of human experience is a marvelous story, and one that AI will change forever. Your story and mine are woven into a fabric along with everyone who came before us, and everyone who will come after. The invitation is to step back for a moment, admire the tapestry, and to see your place within it.

I humbly thank each of you for reading or listening. It means more to me and my family than you could know. I hope you **find peace** wherever you are in the world today.

S. Clay Turner

https://sclayturner.com

https://linkedin.com/in/sclayturner

https://www.youtube.com/@sclayturner

https://medium.com/@sclayturner

https://soaringtitan.com

Getting Oriented

Understanding Intelligent Machines

I realize that I likely have a mixed audience of readers. I expect that some will be students, young professionals, and more established knowledge workers (old people) like myself who are now trying to understand just what happened and how. Honestly, AI took me by surprise. It was not my field. I've always had a deep respect for Data Science, but definitely regarded it as something quite different from my experience as a software engineer. Now, these two worlds have collided and forever merged. It's actually quite awesome. I'd be remiss if I didn't take a step back, however, and try to explain how we really got here. It's been a valuable exercise for me to research and I'm excited to share it with you. Don't worry… I promise that you don't need a background in Computer Science to follow the extraordinary history we'll cover in three, brief chapters.

The Origins of Intelligent Machines

When I was a toddler, I remember staring into the glow of vacuum tubes and hearing strange voices of varied quality and mother tongue, interrupting each other in irregular patterns that seemed otherworldly to the imagination of a small child. Some talked to each other. Some seemed to be curiously unaware of the other voices that were clear to me. Yet even those who seemed so close were oddly oblivious to me and my eavesdropping. Dad was an amateur radio enthusiast and had hot, humming equipment from every decade of the 20th century leading up to my arrival. Just two decades earlier, scientists looked enthusiastically into the same glow and uttered the first pairing of words that we now see in everything, everywhere, and seemingly all at once. But whispers of Artificial Intelligence echo from a time long before Dad's radios and industry's early computers. They were contemplated around the glow, not of vacuum tubes, but of fires, torches, and later candles. As we explore these ancient roots, it's clear that our ancestors wrestled with the same questions occupying our minds today: Where do my thoughts come from? What makes us intelligent? Can we create beings that think as we do?

In the sun-baked streets of Ancient Greece, for example, storytellers spoke of Talos, a bronze giant crafted by the divine smith Hephaestus to guard the shores of Crete. This wasn't merely a tale of fantasy – it was an expression of humanity's deepest aspirations. While their tools were bronze and stone rather than silicon and code, the Greeks imagined possibilities that we're only now beginning to realize. Their myths spoke to something fundamental in the human spirit: our desire to create beings in our own image, to understand the spark of consciousness that makes us who we are.

As civilization marched forward into the Age of Reason, these mythological dreams began to take on more systematic forms. The 17th century brought a profound shift in how we viewed the human mind itself. René Descartes, writing in a time of religious upheaval and scientific revolution, proposed a radical division between mind and body. Picture him in his study, watching animals in the garden below, wondering if their behaviors could be explained as purely mechanical processes. Though he held firm to the uniqueness of human rational thought, his ideas opened a door that others would eagerly walk through.

Thomas Hobbes, living through the tumult of the English Civil War, took an even bolder stance. In an era where traditional authorities were being questioned, he suggested that human thought itself might be a form of computation – a mechanical process that could be understood and potentially replicated. It was a revolutionary idea for its time, one that would lay the groundwork for modern Artificial Intelligence.

Then there was Gottfried Wilhelm Leibniz, a relatively unsung hero in his own time. He would describe the binary language that computers use today, co-author calculus, and posit that all of human reasoning could be reduced to systematic rules. His hope was to one day resolve all matters of debate with logical computation.

These weren't just abstract philosophical debates. They reflected the technological and social transformations of their eras. Just as we grapple with AI's implications in our own time, these thinkers were responding to the mechanical marvels of their day – from intricate clockwork to early calculating machines.

What's fun to see in this history is how it mirrors our own struggles with Artificial Intelligence today. From ancient myths to Enlightenment philosophy, we see the same pattern: human imagination pushing against the boundaries of what seems possible. These early thinkers may not have had our tools, but they shared our fundamental questions: Can machines think? What separates human intelligence from mechanical processes? Where does consciousness truly reside?

In the 100 years leading up to my own birth in 1977, philosophical speculation would begin to give way to practical reality, but these same questions would just gain new urgency. The path from Talos to the Industrial Revolution and on to the century of my birth is not just a timeline of technological progress; it's a story of human aspiration; of our persistent drive to understand and recreate the mysteries of mind and consciousness. It is, of course, a varied history of both light and darkness — beauty and ugliness.

In the shadow of World War II, as humanity grappled with both its destructive potential and its capacity for innovation, a small group of visionaries began to imagine something extraordinary — machines that actually could think. The late 19th and early 20th centuries had already witnessed remarkable transformations: the rise of industrial automation, the birth of quantum physics, and technological advances that would have seemed like magic to previous generations. Yet these dreamers looked beyond the mechanical marvels of their time, pursuing an even more ambitious vision.

At the heart of this transformation stood Alan Turing, a brilliant mathematician whose imagination would help reshape our understanding of what machines could achieve. In 1936, while most of the world was still using mechanical calculators, Turing envisioned something far more profound — a universal machine that could compute anything that could be described through step-by-step instructions. Picture, if you will, a young mathematician at Cambridge, walking along the Cam River, contemplating how to break down human thought processes into their most fundamental components. His theoretical "Turing machine" wasn't just another calculator; it was a window into the very nature of computation itself.

What makes Turing's vision particularly compelling was its convergence with another great mind of the era, Alonzo Church. Working independently, both arrived at the same profound insight - that there were fundamental patterns to computation, universal truths that transcended any particular machine or method. Their combined insight, known as the Church-Turing Thesis, suggested something remarkable: if human reasoning could be broken down into clear steps, a machine might one day replicate it.

But Turing wasn't content with pure theory. As war clouds gathered over Europe, he found himself at Bletchley Park, applying his ideas to break Nazi encryption codes. Here, abstract mathematical concepts became life-saving innovations. The machines he helped design demonstrated that algorithmic thinking could solve real-world problems of staggering complexity. This practical success only fueled his imagination further.

By 1950, in a paper that would become legendary, Turing posed a question that still challenges us today: Could a machine's behavior become indistinguishable from a human's? His proposed test, initially called the "Imitation Game," wasn't just about machines fooling humans — it was about understanding the very nature of thinking itself. What does it mean to think? To reason? To be intelligent? These questions, born from Turing's imaginative leap, still echo through our discussions of AI today.

During this same era, another remarkable figure, John von Neumann, was pushing the boundaries of what we thought possible. While his contributions to computer architecture would become foundational (indeed, most computers today still follow his basic design), it was his work on self-replicating systems that truly captured the imagination. Von Neumann showed how simple rules, properly arranged, could create systems capable of copying themselves — a concept that eerily mimicked biological processes. This wasn't just mathematics; it was a glimpse into how complexity could emerge from simplicity, how intelligence might arise from basic principles.

These pioneers - Turing, von Neumann, and their contemporaries - weren't just building theoretical frameworks; they were dreaming of possibilities that exceeded the technological limitations of their time. They laid the groundwork for what we now call Artificial Intelligence, not through hardware or software (as we know it today), but through the power of human imagination and theoretical models. They dared to ask: Could machines think? Could they learn? Could they evolve?

The Dartmouth Conference (1956)

In the summer of 1956, as the world was still processing the aftermath of World War II and entering an era of unprecedented technological advancement, a small group of visionaries gathered on the serene campus of Dartmouth College in Hanover, New Hampshire. The setting itself spoke to the moment - an institution of learning nestled in New England's rolling hills would become the birthplace of a field that would one day transform human society. While families across America were embracing the prosperity of the 1950s, purchasing their first television sets and witnessing the dawn of commercial computing, these pioneers were imagining something far more ambitious: machines that could think.

The workshop's organizers - John McCarthy, Marvin Minsky, Nathan Rochester, and Claude Shannon — brought together minds from diverse backgrounds, each carrying their own perspective on what it meant to replicate human thought in mechanical form. McCarthy, a young mathematician whose imagination had been captured by the possibility of thinking machines, gave this nascent field its name: "Artificial Intelligence." The term itself was a bold declaration, suggesting that humanity might create something akin to its own intellectual capabilities — an audacious notion that would both inspire and challenge researchers for generations to come.

The gathering came at a pivotal moment in human history. The first electronic computers, though primitive by today's standards, were demonstrating capabilities that seemed almost magical to the average person. These early machines, filling entire rooms with their vacuum tubes and switches, were already solving complex mathematical problems faster than any human could. For the workshop's attendees, this was just the beginning. They saw beyond the basic calculations these machines could perform, envisioning systems that could reason, learn, and perhaps even understand the world as humans do.

Their optimism was remarkable, even infectious. Many believed they were on the cusp of unlocking the mysteries of human cognition, that within a single generation, machines might rival human intellectual capabilities. They approached this challenge through the lens of symbolic logic — the idea that human thought could be broken down into manipulable symbols and rules. This approach, while seemingly simplistic from our modern perspective, laid the groundwork for decades of research and development.

The impact of this modest gathering rippled through time, leading to the establishment of AI laboratories at prestigious institutions like Stanford, MIT, and Carnegie Mellon. Early programs emerged that could prove mathematical theorems and solve puzzles, tasks once thought to require human intelligence. While these achievements were limited in scope, they represented humanity's first steps toward creating machines that could think, or at least simulate thinking in meaningful ways.

Looking back from our vantage point today, where AI systems can engage in natural conversations, generate art, and solve complex problems across countless domains, the Dartmouth Conference might seem like a quaint historical footnote. Yet the questions raised during that summer month in 1956 remain relevant today. As we grapple with AI systems that increasingly match or exceed human capabilities in specific domains, we're still exploring the same fundamental questions about the nature of intelligence, the relationship between human and machine cognition, and the boundaries of what we can create through technology and imagination.

Dartmouth Conference Key Figures and Contributions

Pioneer	Contributions
John McCarthy	Coined the term "Artificial Intelligence" and developed the Lisp programming language, which became standard for AI work. He pioneered concepts like time-sharing on computers, anticipating interactive computing.
Marvin Minsky	Mathematician who transitioned to cognitive science. He conducted significant work in robotics, neural networks, and symbolic processing. Along with McCarthy, he co-founded the MIT AI Lab, which became a major center for AI research.
Nathan Rochester	IBM researcher who helped shape the conference's direction. He played a crucial role in connecting AI's aspirational goals with emerging computer engineering.
Claude Shannon	Known as the "father of information theory," Shannon established foundational work in binary logic and communication. He helped demonstrate AI's potential to unite various branches of mathematical and computational thought.

First Wave: Symbolic AI (1950s—1970s)

The 1950s marked the dawn of what we now call the First Wave of Artificial Intelligence, or "Symbolic AI." These early pioneers, emerging from a world forever changed by the mathematical and technological breakthroughs of wartime computing, dared to ask a profound question: Can we teach machines to think the way we do?

Their approach, later affectionately dubbed "Good Old-Fashioned AI" (GOFAI), reflected both the optimism and the methodical thinking of the post-war era. Just as the world was being rebuilt through careful planning and systematic effort, these researchers believed they could construct Artificial Intelligence piece by piece, rule by rule. Their vision was compelling in its simplicity: human intelligence, they theorized, could be recreated by teaching machines to manipulate symbols according to logical rules – much like how we use language and mathematics to make sense of our world.

Figures like John McCarthy and Marvin Minsky shaped the field's early direction with an unwavering faith in the power of logical reasoning. They envisioned intelligence as a kind of mathematical dance, where each step followed inevitably from the last according to precise rules. Their systems, built on "if-then" logic, were like intricate clockwork mechanisms of thought

— beautiful in their precision, yet ultimately revealing how different machine reasoning was from the fluid, intuitive way humans actually think.

The practical manifestations of this approach were remarkable for their time. Expert systems emerged that could diagnose diseases and analyze molecular structures. DENDRAL, born in the 1960s, became a trusted advisor to chemists, while MYCIN demonstrated an almost human-like ability to diagnose bacterial infections. Perhaps most captivating was ELIZA, created by Joseph Weizenbaum in 1966, which simulated a psychotherapist through clever pattern matching — something that pointed to the Natural Language Processing (NLP) we see today. Though simple by today's standards, ELIZA sparked profound questions about the nature of human-machine interaction that we're still grappling with today.

In the realm of games, which had long served as a measure of human intellectual prowess, machines began making their first tentative steps toward competency. Arthur Samuel's Checkers Player and Richard Greenblatt's Mac Hack chess program demonstrated that computers could engage in strategic thinking, even if their approach was more brute force calculation than human-like intuition. These early achievements, though modest by today's standards, were like the first hesitant steps of a child – promising glimpses of capabilities yet to come.

Yet for all their achievements, these early systems also illuminated the vast complexity of human intelligence. When faced with tasks that humans find effortless — recognizing a friend's face, understanding a joke, or making sense of a chaotic world with incomplete information — these logical, rule-based systems often fell short. This limitation was not a failure of imagination or effort, but rather a profound insight into the nature of intelligence itself. Human cognition, it seemed, wasn't just about following rules; it was about something far more mysterious and beautiful. Like a child's drawing that captures the essence but misses the nuance of reality, these first attempts at machine intelligence achieved remarkable feats while simultaneously highlighting how much we had yet to learn.

Perhaps the most revealing challenge emerged from what researchers came to call the "knowledge bottleneck." Creating these early systems required an almost Herculean effort of translation. In other words, human experts had to meticulously convert their intuitive understanding into explicit rules that machines could follow. Imagine trying to teach someone to ride a bicycle by writing down every minute adjustment of balance and pressure. Just as that task quickly becomes overwhelming, these early AI systems struggled under the weight of increasingly complex rule sets needed to handle real-world situations.

The challenge wasn't just in capturing knowledge, but in scaling it. Each new domain required its own carefully crafted set of rules and exceptions. What worked brilliantly for analyzing molecular structures in DENDRAL couldn't help MYCIN diagnose infections, and neither system could help a computer understand a simple joke or recognize a face. It was as if each system spoke its own highly specialized language, unable to generalize or adapt beyond its narrow domain of expertise.

Most tellingly, these systems struggled with the kind of ambiguity that humans navigate effortlessly every day. While a doctor can make informed decisions with partial information, drawing on intuition and experience, these early AI systems required complete and consistent data to function. They were like brilliant but literal-minded students who could solve complex equations but became paralyzed when asked to interpret a poem.

The contrast between the field's early optimism and these emerging realities was stark. The dreams voiced at the historic 1956 Dartmouth Conference began to collide with technical limitations that proved far more challenging than anyone had anticipated. It was becoming clear that human intelligence wasn't just a matter of following rules, no matter how sophisticated those rules might be.

These limitations weren't failures, but rather invaluable lessons that would shape the field's evolution. Just as early attempts at human flight taught us that mimicking birds' wings wasn't the answer, these first approaches to Artificial Intelligence revealed that mimicking human logical reasoning alone wouldn't be sufficient. The path forward would require new paradigms, different approaches, and a deeper understanding of both human and machine intelligence. As often happens, however, the limitations we face in the present will shape our thinking and curb our enthusiasm.

AI Winter (1970s & 1980s)

The late 1960s marked a sobering transition in Artificial Intelligence research. Like many profound human endeavors, AI's early years were characterized by boundless optimism — a reflection of the post-war era's technological achievements and seemingly limitless potential. As we saw, researchers in the 1950s and early 1960s had painted vivid pictures of machines that could think, reason, and even understand human languages as naturally as we do. Their imagination sparked a wave of excitement that attracted both government funding and private investment. Yet as the 1970s approached, the gap between these dreams and reality began to widen, setting the stage for what would become known as the "AI Winter."

This period of disillusionment offers us a powerful lesson about the relationship between human imagination and technological progress. In the United Kingdom, the pivotal Lighthill Report of 1973 cast a harsh light on AI's practical limitations. Professor James Lighthill's critical evaluation led to dramatic funding cuts, forcing many British universities to redirect their AI research efforts. Across the Atlantic, even DARPA — one of AI's most steadfast supporters — began to question its investments, particularly as projects in Natural Language Processing and robotics struggled to deliver on their promises.

The challenges weren't purely financial or theoretical. The machines of this era, though remarkable for their time, simply couldn't keep pace with the researchers' ambitions. Imagine trying to build a skyscraper with hand tools — the vision might be clear, but the infrastructure simply wasn't ready. Computer memory was precious and expensive. Processing power was so limited that operations we now complete in seconds could take hours or days. Perhaps most critically, researchers lacked the vast pools of data that modern AI systems use to learn and improve.

Some commercial successes did emerge, particularly in narrow applications. Expert systems like MYCIN and PROSPECTOR showed promise in specific domains like medical diagnosis and geological analysis. But these systems were brittle, failing when pushed beyond their carefully defined boundaries. They required painstaking work from human experts to encode every rule and relationship — a process that proved too costly and time-consuming for widespread adoption.

As funding dwindled and labs closed, the AI research community adapted and evolved. Rather than pursuing the dream of general Artificial Intelligence, many researchers shifted their focus to more specialized problems. Some turned to cognitive science, seeking deeper insights into human intelligence that might one day inform Machine Learning. Others laid the groundwork for practical applications in robotics and computer graphics. This period of constraint, though challenging, fostered a culture of incremental progress that would prove vital in the decades to come.

So even in these quiet years, the seeds of future breakthroughs were being planted. Early versions of neural networks, decision trees, and Bayesian frameworks took shape during this winter. While data remained scarce and computing power limited, researchers continued to develop the theoretical foundations that would later revolutionize the field. Moore's Law pressed forward relentlessly, and by the late 1980s, the cost of computing and storage began to fall, setting the stage for AI's renaissance in the 1990s.

The story of the AI Winter reminds us that technological progress rarely follows a straight line. Like the Wright brothers testing countless wing designs before achieving first flight, or Edison testing thousands of materials for his light bulb, the path to Artificial Intelligence has been marked by setbacks that ultimately strengthened the field. Today's AI researchers and developers stand on the shoulders of those who persevered through these challenging decades, their work quietly laying the foundation for the breakthroughs we now witness. Indeed, we all "stand on the shoulders of giants."

As we marvel at today's rapid advances in AI, the lessons of this period remain relevant. They remind us to balance our imagination with practical constraints, to value incremental progress alongside revolutionary ambition, and to recognize that even periods of apparent dormancy can nurture profound innovation. The cycle of hype, disillusionment, and renewed progress continues to shape our relationship with technology, teaching us that true advancement often comes not from avoiding winter altogether, but from learning to grow stronger through its challenges.

Second Wave: Statistical Learning (1980s—2000s)

As the 1980s dawned, a profound shift was occurring in how we imagined machines could think. The earlier dreams of creating intelligence through carefully crafted rules and logic were giving way to a new vision - one where machines could learn from experience, much like humans do. This transition marked what we now call the Second Wave of AI, and it emerged during a time of remarkable transformation in human society.

The 1980s — my favorite decade as a Gen-X'r — were defined by rapid technological advancement and cultural change. Personal computers were finding their way into homes and offices. The Berlin Wall would fall by decade's end. A new generation was growing up with video games and the early promises of digital connectivity. Against this backdrop of transformation, AI researchers were questioning their fundamental assumptions about machine intelligence.

The previous decades had shown us that programming explicit rules for every situation was like trying to teach a child by having them memorize every possible scenario they might encounter. It simply couldn't scale. Instead, researchers began to imagine a different approach, one that would let machines discover patterns on their own through data and statistical analysis. This shift wasn't just technical; it represented a fundamental change in how we thought about intelligence itself.

As computing power grew steadily through the 1980s and 1990s, following the reliable march of Moore's Law, new possibilities emerged. Researchers could now work with larger datasets than ever before, though still tiny by today's standards. The UCI Machine Learning Repository became a treasure trove for algorithm development, while research labs collected speech samples, images, and text that would help machines learn to see, hear, and understand language.

The real-world impact of these advances began to touch everyday lives. By the late 1990s, you could speak to your computer using Dragon NaturallySpeaking, though with mixed results. Banks started using statistical algorithms to detect fraudulent transactions in real-time. Early e-commerce sites experimented with recommending products based on customer behavior. These applications, while primitive by today's standards, showed us glimpses of a future where machines could adapt and learn from experience.

Perhaps the most emblematic moment of this era came in 1997, when IBM's Deep Blue defeated world chess champion Garry Kasparov. Though Deep Blue relied more on raw computational power than statistical learning, it symbolized humanity's growing ability to create machines that could match and even exceed human capabilities in specific domains.

The Internet's rapid growth through the 1990s brought another dimension to this evolution. As more human activity moved online, we began generating digital footprints that would later fuel the next wave of AI advancement. While we didn't yet have the massive datasets of today's social media era, the groundwork was being laid for what would become the Big Data revolution.

In research labs and universities, new theoretical frameworks emerged. Support Vector Machines (SVMs) provided mathematical rigor to the field of Machine Learning. Statistical approaches to language translation showed that machines could begin to bridge the gaps between human languages not through rules, but through patterns found in millions of translated sentences.

By the early 2000s, this Second Wave had demonstrated something profound: machines could indeed learn from experience, finding meaningful patterns in real-world data. While we still needed to carefully guide their learning — selecting features and representations that would help them understand the world — we had moved beyond purely rule-based systems. The stage was set for the next great leap forward, when neural networks would return to center stage and usher in the era of Deep Learning.

Third Wave: Deep Learning Boom (2010s — early 2020s)

As we entered the 2010s, the world was experiencing a profound digital transformation. Social media had connected billions of people, smartphones were becoming ubiquitous, and the Internet was weaving itself into the fabric of daily life. Against this backdrop of digital revolution, AI researchers began to unlock the potential of an approach that had laid dormant for decades: neural networks. This renaissance would come to be known as the Deep Learning Boom, though its seeds were planted long before its flowers would bloom.

The story of this third wave of AI is, at its heart, a tale of convergence. Just as a perfect storm requires specific conditions, Deep Learning needed three elements to align: vast amounts of digital data, unprecedented computational power, and human ingenuity to bring it all together. The explosive growth of social media, e-commerce, and streaming services provided the data. The gaming industry's insatiable appetite for better graphics had inadvertently created the perfect computers for AI in the form of Graphics Processing Units (GPUs). All that remained was for researchers to reimagine how these pieces could work together.

The breakthrough moment came in 2012, when researchers achieved something remarkable in computer vision. Using an architecture called AlexNet, they taught computers to "see" in ways that seemed impossible just months before. This wasn't just an incremental improvement — it was a quantum leap that caught the attention of the entire technology world. The approach, known as Convolutional Neural Networks (CNNs), mimicked the human visual cortex in fascinating ways. Early layers of the network would detect simple edges and shapes, just as our own visual system does, while deeper layers would gradually build up to recognize complex patterns and objects.

As remarkable as this achievement was, it was just the beginning. Researchers soon discovered that similar principles could be applied to language and speech. They developed architectures called Recurrent Neural Networks (RNNs) that could maintain a kind of digital memory, allowing them to understand context in ways that previous systems could not. By 2017, a new architecture called the Transformer would emerge, revolutionizing how machines process language. This innovation would eventually lead to systems like GPT and BERT, which would bring AI into mainstream consciousness and spark global conversations about the future of human-machine interaction.

In many ways, the Deep Learning Boom mirrored broader transformations happening in society. Just as social media was reshaping how we connect, and smartphones were changing how we interact with the world, Artificial Intelligence was quietly transforming the very foundation of how machines learn and operate. Like a child discovering their senses for the first time, AI systems were beginning to see, hear, and understand the world in ways that seemed like science fiction just years before.

The scale of this transformation was made possible by an extraordinary convergence of technological abundance. We found ourselves awash in digital data — every click, every photo, every message becoming part of a vast digital tapestry that could train these hungry learning systems. In an almost poetic twist, the same graphics cards that helped gamers explore virtual worlds became the engines of AI innovation. These GPUs, with their ability to perform countless calculations in parallel, turned out to be perfectly suited for training neural networks. It was as if we had been unknowingly building the infrastructure for an AI revolution all along.

But perhaps most remarkable was the democratization of AI that followed. What was once the exclusive domain of research laboratories and tech giants became accessible to anyone with a laptop and an Internet connection. Open-source tools like TensorFlow and PyTorch emerged, turning complex mathematical concepts into building blocks that developers, like me, could assemble like digital Lego pieces. The global AI community grew, and with it, a culture of sharing and collaboration that accelerated progress beyond what any single organization could achieve alone.

The real-world impact of these advances began to touch every corner of society. In medicine, AI systems learned to spot tumors in medical images with remarkable accuracy, becoming tireless assistants to radiologists. In automotive technology, cars began to see and understand their surroundings, taking the first steps toward autonomous driving. Virtual assistants like Alexa and Google Assistant brought AI into our homes, while recommendation systems on platforms like Netflix and Spotify began to understand our preferences with almost unsettling precision.

Yet as these systems grew more capable, they also held up a mirror to our own biases and limitations. AI systems trained on historical data began to reflect and amplify societal prejudices, leading to discriminatory outcomes in everything from hiring tools and lending decisions, to inadvertently running over human beings not recognized as human beings. The same technology that could create beautiful art or write compelling prose could also generate deepfakes and spread misinformation. We found ourselves grappling with fundamental questions about privacy, consent, and the very nature of human creativity and work.

By the early 2020s, as ChatGPT captured the world's imagination and brought AI into everyday conversation, it became clear that we were standing at a pivotal moment in human history. The tools we had created were no longer just aids to human intelligence, but were beginning to exhibit capabilities that challenged our understanding of what machines could do. As policymakers and global agencies scrambled to establish frameworks for this new reality, many of us found ourselves contemplating deeper questions: What does it mean to be uniquely human in an age of thinking machines? How do we preserve our essential humanity while harnessing these powerful tools?

These questions have become increasingly urgent as we moved into the mid-2020s, and they are no longer just philosophical musings. They have become practical concerns that will shape how we live, work, and relate to one another in a world where the line between human and machine intelligence is growing ever more blurred. The Deep Learning Boom did not just transform technology – it has fundamentally altered our relationship with machines and with each other.

The Current Status of Intelligent Machines

Standing now in early 2025, we find ourselves immersed in a world where Artificial Intelligence has transcended its academic origins to become an intimate part of our daily existence. Like electricity before it, AI has woven itself so thoroughly into the fabric of our lives that we often forget it's there — quietly suggesting our next song, anticipating our grocery needs, or helping us navigate through traffic. Yet unlike electricity, these systems don't just power our appliances; they think alongside us, reason with us, and increasingly, "create" with us. In 2025, we're witnessing not just a technological revolution, but a reimagining of human potential. The "quote-unquote" tools we've created are beginning to think alongside us, and we are now starting to hand our tools over to them, giving them "agency."

In our homes, AI assistants like Siri and Alexa have evolved beyond simple command-takers into ambient presences that feel almost like family members. They manage our schedules, control our environments, and even make calls on our behalf. It's a peculiar intimacy we've developed with these digital companions, one that would have seemed strange just a decade ago but now feels as natural as checking our phones in the morning.

In the workplace, AI has become the quiet colleague who never sleeps, processing vast streams of data to help make decisions that once relied purely on human intuition. Banks use it to detect fraud in real-time, doctors leverage it to spot patterns in medical images that human eyes might miss, and manufacturers employ it to predict equipment failures before they happen. Predictive analytics now guide decisions worth millions of dollars, while robotic process automation handles countless routine tasks that once occupied human hours. Manufacturing floors hum with smart systems that can spot microscopic defects invisible to human eyes, while supply chains restructure themselves in real-time based on AI-powered insights. These aren't just improvements in efficiency – they represent a fundamental shift in how we organize human work and creativity. Most of these systems are intended to amplify human expertise, allowing us to focus on the uniquely human aspects of our work — creativity, empathy, and complex problem-solving.

Perhaps most fascinating is how AI has begun to reshape creative expression itself. Writers now collaborate with language models that can help refine their prose or suggest new directions for their narratives. I am no exception. Artists partner with image generation systems that can transform written descriptions into visual art in seconds. Software developers work alongside AI coding assistants that can understand their intentions and help bring them to life in code. This reality shapes perhaps as much as 80 to 90 percent of my code writing now as a technical founder and CTO. AI like this doesn't have to replace human creativity or expertise; it can expand the boundaries of what we can imagine and create. In fact, there is a growing catch phrase I'm hearing in 2025: **"Those not being replaced by AI are those using AI."** I think this will increasingly become an axiom.

What makes this moment in history particularly remarkable is not just the capabilities of these systems, but how quickly they've become normalized. Children growing up today will never know a world without AI companions. For them, the ability to have a natural conversation with a machine or to generate art from text will be as unremarkable as color television was to my generation. This mainstreaming of AI brings with it profound questions about our relationship with technology. As recommendation engines on platforms like YouTube and Netflix learn our preferences with increasing precision, we have to consider how these systems shape our worldview. When our entertainment, news, and social connections are all filtered through AI-driven personalization, what happens to serendipity? To discovery? To the beautiful accidents that often lead to creativity, empathy, and complex problem-solving?

APIs for AI Models

The barriers to entry for leveraging AI have dropped so dramatically that organizations of any size can now incorporate AI capabilities that rival those of tech giants. A startup can integrate sophisticated Natural Language Processing without a dedicated AI team. A solo developer can create applications that leverage the same image generation capabilities as major design platforms. Gone are the days when leveraging AI meant training models from scratch or maintaining complex infrastructure. Today's landscape offers a rich tapestry of API endpoints that seamlessly connect our applications to some of the most sophisticated models ever created.

The breadth of these offerings is staggering. Text generation APIs have evolved far beyond simple completion tasks, now offering capabilities that span from nuanced conversation to complex reasoning. Code generation APIs have transformed development itself, with models that don't just complete functions but understand architectural patterns, best practices, and cite their inspiration. Image generation APIs have progressed from creating basic variations to producing photorealistic imagery from natural language descriptions. Video generation, once the stuff of science fiction, has become accessible through simple API calls that can create dynamic content from text prompts. Voice synthesis and speech recognition APIs have achieved such fidelity that the line between human and machine-generated speech has become remarkably thin.

What makes this truly transformative is the computational offloading these APIs enable. When I think back to my early SaaS days, I remember the constant balance of infrastructure costs against capabilities. Today's developers face a different reality. The most computationally intensive tasks – *training and running large language models, generating complex images, or processing video* – can be delegated to robust cloud services through simple API calls. This isn't just convenience; it's a fundamental shift in how we approach application development. With open-source models becoming more powerful and aligned to the capabilities of closed-source, frontier models, even self-hosting with cloud services becomes more achievable and defendable at a certain scale.

The integration patterns are where things get interesting. Developers are creating sophisticated workflows by chaining these APIs together. Imagine an application that takes a user's voice input, transcribes it through a speech recognition API, processes the intent through a language model, generates corresponding images or code, and responds with synthesized speech — all orchestrated through a series of API calls. This kind of integration would have seemed overwhelming just a few years ago. Now it's becoming commonplace with frameworks like LangChain and LlamaIndex.

This accessibility is particularly meaningful as we approach the era of AI agents. These APIs aren't just services we call; they're becoming the building blocks of autonomous systems that can work independently to accomplish complex tasks. The implications for developers, businesses, and society at large are profound. We're not just democratizing access to AI; we're creating the foundation for a new kind of software development where human creativity and machine intelligence can truly amplify each other — or "compete" with one another.

Notable AI APIs By Category (o3-mini w/ Deep Research, February 2025)

Category	Key Providers & Models	Notable Updates & Trends	Common Applications
Text Generation & NLP	<ul><li>OpenAI (o3, o1, GPT-4)</li><li>Anthropic (Claude 3.5)</li><li>AI21 (Jurassic-2)</li><li>DeepSeek (R1, V3)</li><li>Cohere</li><li>Google (PaLM 2, Gemini)</li><li>Meta (LLaMA 2)</li><li>Amazon (Bedrock/Titan)</li><li>Mistral AI</li><li>Aleph Alpha (Luminous)</li><li>MosaicML/Databricks (MPT)</li></ul>	New open-source and commercial LLMs beyond OpenAI (e.g. LLaMA 2, Gemini). Some older models deprecated (e.g. Codex, GPT-3). Trend toward larger context (100k+ tokens) and multimodal inputs.	Chatbots & virtual assistants; content generation (articles, marketing copy); summarization and research Q&A; translation & localization; writing and code assistance.
Code Generation	<ul><li>GitHub Copilot (OpenAI GPT-4 backend)</li><li>Amazon CodeWhisperer</li><li>Tabnine</li></ul>	Codex API shut down (2023); Code Llama open-sourced (free use) improving state-of-art open code AI; IDE	Autocomplete & pair programming in IDEs; generating functions from descriptions; explaining and

	• Meta Code Llama • BigCode StarCoder • Replit Ghostwriter • Google Codey (PaLM) • Cursor	integration ubiquitous (VS Code, IntelliJ, etc. all have AI). New entrants like StarCoder (15B open model) provide alternatives.	debugging code; writing tests and documentation; accelerating software development cycle.
Image Generation	• OpenAI DALL·E 3 • Midjourney v5 • Stability AI (Stable Diffusion & SDXL via DreamStudio) • Adobe Firefly • Canva Magic Media • Bing Image Creator • Ideogram • Leonardo AI	DALL·E 3 (2023) improved prompt understanding and is integrated into ChatGPT/Bing. Midjourney remains top for photorealism. Stable Diffusion open model enables countless custom derivatives. Adobe Firefly focuses on commercially safe data. New tools address specific gaps (e.g. Ideogram for legible text in images). Growing focus on ethics (licensed training data, watermarks).	Graphic design and marketing content (ads, social posts); concept art and creative design; virtual photography for e-commerce (product images, room decor visuals); story illustration and game asset creation; personalized art and avatars for users.
Video Generation	• Runway ML (Gen-2, Gen-3 'Act One') • Synthesia • D-ID • HeyGen • Pika Labs • DeepBrain/Colossyan (AI avatars) • NVIDIA & Google research (Imagen Video, Phenaki)	Text-to-video now possible in short clips (Runway Gen-2); improvements in consistency and length (Runway's latest can maintain characters over multiple scenes). AI avatar videos (Synthesia, D-ID) nearly photoreal for speech presentations. Early multi-scene storytelling with AI is emerging (Pika, Runway "Story Mode"). Increased funding into startups (e.g. Synthesia $90M+) and features like multi-language avatar dubbing.	Corporate training and how-to videos with AI presenters; marketing videos and personalized video messages; prototype film scenes and game cinematics; social media content (short AI-generated skits, dynamic visuals); automating video localization (AI dubbing in various languages).
Speech & Audio	• ElevenLabs (prime quality cloning) • Play.ht • Resemble AI • Microsoft Azure TTS • Google Cloud TTS • Amazon Polly • Murf.ai • WellSaid Labs • Meta (AudioCraft research)	Ultra-realistic voice cloning widely available (ElevenLabs et al.) – raising need for deepfake detection. Multilingual TTS in single voices (cross-language) now common. Big tech TTS (Google, Microsoft) offer 200+ voices and custom voice training. New models generate music and SFX (Meta's AudioCraft, Google MusicLM). Voice assistants now use more emotive, human-like tones.	Audiobook and article narration; call center IVRs and digital assistants with natural voices; game and animation voiceovers without actors; voice content localization (dubbing) with voice preservation; accessibility (screen readers, personal voice for speech-impaired users); podcast production and media voiceover.
Multi-Modal AI	• OpenAI GPT-4V (Vision) • Anthropic Claude 3.5 • Google Gemini • Microsoft Kosmos-1 • Meta ImageBind &	Vision-and-language models deployed (you can send images to GPT-4 and Claude 3 now and get analysis). Gemini trained from scratch	Visual Q&A and analysis (describe image, interpret chart, troubleshoot from a photo); digital assistants that see and hear (e.g. AR

	SeamlessM4T • HuggingFace Multimodal pipelines	on text+images+audio – setting new SOTA on multi-modal reasoning. Assistants can converse, see, and speak (ChatGPT with voice and vision, Bard with image understanding). Early multimodal agents can take actions (clicking buttons, etc.) by interpreting visual context. Integration of modalities is a major research focus, moving towards unified models that handle text, vision, and audio together.	assistants guiding repairs or cooking); medical and scientific data analysis (combining lab results with scans); creative content generation loops (using one modality's output as another's input – image to story, sketch to image to animation); more natural human-AI interaction where you can show and tell, not just tell.

AI Agents and Agentic Processing

As we enter 2025, tech giants are acknowledging that SaaS will give way to AI agents as the primary force transforming how we interact with technology and accomplish work. These sophisticated entities mark a shift from rigid, predefined systems to adaptive, autonomous ones that can reason, learn, and evolve through experience.

Modern AI agents are complex software programs that combine cutting-edge Machine Learning – *typically through a specified Large Language Model via API* – with traditional rules that define their problem domain. What distinguishes them from conventional software is their ability to operate independently within defined parameters, making decisions and taking actions without constant human oversight. This autonomy comes from sophisticated architecture that enables them to process and learn from various inputs, whether interpreting natural language, analyzing sensor data, or parsing vast amounts of information.

The true power of AI agents lies in their adaptability and decision-making capabilities. Unlike traditional software following linear if-then statements, these agents dynamically adjust their approach based on new information and past experiences. They learn not only from interactions with users and environments but also from other AI agents, creating a network effect that enhances their collective capabilities. Crucially, agents can use electronic tools like we do, e.g., Internet browsers and APIs, that allow them to act on their decisions to accomplish real work.

We're seeing early practical applications across industries. In corporate settings, scheduling agents are transforming meeting coordination by navigating multiple calendars, time zones, and availability patterns to propose optimal meeting times. They learn from past scheduling patterns and participant preferences to make increasingly refined suggestions. In market research, AI agents serve as digital analysts, continuously monitoring Internet sources to identify meaningful trends and patterns, compiling comprehensive reports that synthesize diverse data points. In manufacturing, specialized agents will monitor complex machinery in real-time, analyzing performance metrics to predict potential issues before they become critical and automatically coordinating preventive maintenance.

Agent Frameworks

As AI capabilities expanded rapidly in 2023, developers began creating frameworks that would allow AI systems to work more independently and collaboratively. These frameworks, which we now call "agentic," emerged as the building blocks for a new kind of digital workforce. They represent a crucial step between the early days of AI assistants that could only respond to direct questions and truly autonomous systems that can understand complex goals, break them down into smaller tasks, and work independently or with other AI agents to achieve them.

LangChain emerged as an early pioneer, giving developers tools to create AI systems that could access external information and follow structured workflows. Microsoft Research then introduced AutoGen, focusing on systems where multiple AI agents could work together, much like human teams collaborating on complex projects. AgentGPT followed, emphasizing autonomous agents that could handle multi-step tasks with minimal human oversight.

As developers began experimenting with these technologies, new frameworks addressed specific needs. CrewAI introduced the concept of role-based teams, where different AI agents could specialize in particular tasks, similar to how human organizations distribute expertise across departments. Langflow made these powerful tools accessible to non-programmers through visual interfaces, with hopes of accelerating adoption across industries that commonly lack technical resources.

The real power of these frameworks lies in their ability to orchestrate complex workflows. Think of them as conductors in an orchestra, coordinating multiple AI agents that each contribute their unique capabilities to achieve a larger goal. These agents can maintain context over time, learning from past interactions and adapting to new situations. They can communicate with each other, sharing insights and delegating tasks, much like human colleagues would.

In practice, this means that tasks that once required careful human coordination — *like researching market trends, analyzing data, or managing complex projects* — can now be handled more efficiently through AI collaboration. The human role shifts from executing every detail to providing strategic direction and creative insight, allowing people to focus on what they do best: imagining possibilities and making nuanced judgments that machines still struggle to replicate.

Agentic vs. AGI: Where They Differ

There's a distinction between AI agents as we know them today — *often described as narrowly specialized "agentic" or "agent like" systems* — and the long-envisioned prospect of Artificial General Intelligence (AGI). Current AI agents demonstrate remarkable capabilities within defined tasks, leveraging specific tools and data to coordinate complex workflows. They can excel at well-defined objectives and exhibit a kind of "agency," but they still rely on narrow, domain-specific strengths.

AGI, by contrast, denotes the aspiration for an artificial intelligence with broad, flexible reasoning akin to a human's — capable of everything from writing poetry to solving abstract mathematical puzzles. The latest generation of Large Language Models have surprised many by showing emergent abilities that cut across disciplines, sparking fresh debate about whether we might be approaching general intelligence. Yet many experts maintain that true AGI, defined as possessing robust, human-like adaptability and understanding, remains an open challenge. Determining how close we are to such a breakthrough is still subject to ongoing research and diverse opinions. The latest "reasoning" models, like OpenAI's o3, o1, and DeepSeek's R1, raise new questions and debate about just how near or far we actually are from achieving AGI.

Benefits of AI Agents and Agentic Workflows

We've all interacted with machines through explicit commands for a very long time — clicking buttons, typing instructions, or following predetermined paths through a UI/UX. Even with today's sophisticated AI models, we typically engage in a back-and-forth dialogue: we ask, they respond, and the cycle continues. But imagine instead having a digital partner that understands your broader intentions and can work independently toward your goals, much like a human assistant would. The rise of AI agents marks a transition from seeing technology as a collection of digital tools we actively use to partners we collaborate with, using their own collection of digital tools. Instead of us adapting to the rigid structures of computer systems, these agents adapt to our natural way of working and thinking. They can understand context, remember past interactions, and most importantly, take initiative to accomplish tasks without constant supervision. Imagine AI agents as skilled workers with access to a comprehensive digital toolbox. Just as a master craftsperson knows exactly which tool to use for each task, these agents can seamlessly select and employ different technologies to accomplish their goals.

In your personal life, consider how you might plan a family vacation today. You spend hours juggling multiple browser tabs, comparing flight prices, reading hotel reviews, checking weather patterns, and coordinating everyone's schedules. Now imagine instead telling an AI agent, "I'd like to plan a two-week European vacation for my family of four this summer, focusing on historical sites and local cuisine." The agent could then work autonomously to research options, consider your past travel preferences, monitor price trends, and gradually build a comprehensive plan — all while you focus on other priorities. When it encounters decisions requiring your input, it asks thoughtful, specific questions rather than overwhelming you with endless options.

In business, think of a modern sales organization. An AI agent might serve as a virtual sales assistant, but its power comes from how it orchestrates multiple tools and systems. It could monitor customer relationship management (CRM) software for new leads, analyze communication patterns in email systems, and track engagement metrics across marketing platforms. When it spots an opportunity, it might automatically draft personalized outreach messages, schedule follow-ups, and even prepare preliminary proposals by pulling relevant case studies from the company's document management system.

In manufacturing environments, agents become sophisticated control centers that interact with an array of industrial systems. They can communicate with robotic process automation (RPA) tools to manage assembly lines, interface with enterprise resource planning (ERP) systems to optimize inventory levels, and leverage computer vision systems to perform quality control inspections. When they detect anomalies or potential improvements, they can automatically adjust production parameters or alert human supervisors with detailed analyses and recommendations.

The financial sector offers particularly compelling examples of tool integration. AI agents can simultaneously monitor multiple data feeds — market trends, economic indicators, company financials, and social media sentiment — while interfacing with trading platforms to execute precisely timed transactions. They can integrate with risk management systems to ensure compliance, communicate with accounting software to reconcile transactions, and generate detailed reports for human review. This level of coordination would be impossible for human operators to maintain consistently.

Project management provides another illuminating example. Imagine an AI agent that serves as a project coordinator, interfacing with team collaboration tools like Slack or Microsoft Teams, project management platforms like Jira or Asana, and documentation systems like Confluence or SharePoint. It can automatically update project timelines based on team progress, redistribute tasks when it detects potential bottlenecks, and even draft status reports by synthesizing information from multiple sources. The agent might also integrate with calendaring systems to schedule meetings at optimal times, taking into account team members' working hours and preferences.

In customer service environments, agents can become sophisticated orchestrators of multiple support channels. They can monitor social media platforms for mentions requiring attention, manage email support queues, handle chat interactions, and even interface with telephony systems. What makes this particularly powerful is their ability to maintain context across all these channels while accessing knowledge bases, ticketing systems, and customer histories to provide consistent, informed responses.

The healthcare sector demonstrates how agents can work with specialized technical tools. An AI agent might integrate with electronic health record systems, medical imaging software, laboratory information systems, and pharmaceutical databases. It could help coordinate patient care by monitoring treatment schedules, checking for drug interactions, and alerting healthcare providers to potential concerns — all while ensuring compliance with privacy regulations and medical protocols.

This shift toward autonomous agents is becoming possible through several key technological advances. Modern AI models can now understand natural language with remarkable sophistication, allowing them to interpret complex instructions and context. They can interface with a wide range of digital systems and APIs, giving them the ability to take real actions in the world — from booking appointments to analyzing documents to controlling smart home devices.

Perhaps most importantly, they're developing the ability to break down complex goals into manageable steps and adapt their approaches based on changing circumstances. Looking ahead, we're seeing the emergence of agents that can even create and modify their own tools through code generation. These agents might write scripts to automate new processes, create custom data analysis tools, or develop specialized interfaces for unique business needs simply because they reason on their own that it's the best way to achieve their goal — much like we might create our own spreadsheets, apps, or utilities. This represents a significant evolution from agents that simply use existing tools to agents that can extend their own capabilities based on emerging requirements.

What makes this truly revolutionary is the potential for agents to learn and improve through interaction. Unlike traditional software that remains static until manually updated, well-engineered AI agents can refine their understanding of your preferences, adapt to your working style, and even anticipate your needs based on observed patterns. They become more valuable partners over time, much like human colleagues who grow more effective as they gain experience. This can make the difference between having several skilled workers and having a coordinated team that instinctively knows how to work together. As these agents become more sophisticated in their ability to leverage different tools and technologies, they will undoubtedly create new possibilities for business efficiency and innovation, enhancing the productivity of human workers, and allowing them to focus on the strategic and creative aspects of their roles that truly require human insight. That's, at least, "the hope."

The Future of Intelligent Machines

From Agentic Systems to General Intelligence

Although we've already introduced it, let's expand a bit on the significance of Artificial General Intelligence, or AGI. While today's AI systems excel at specific tasks like interpreting medical scans or engaging in conversation, they operate within carefully defined boundaries, much like a virtuoso who has mastered a single instrument but cannot conduct the entire orchestra. AGI represents something more profound — a technological leap toward machines that can think, reason, *and learn* across any domain with the same fluidity and adaptability that characterizes human intelligence. Okay, so we've kind of said that before, but what does that really mean?

Picture a child who learns to ride a bicycle and then instinctively applies those same principles of balance to skateboarding. This natural ability to transfer knowledge from one context to another exemplifies the kind of general intelligence we seek to create. An AGI system would demonstrate this same versatility, moving effortlessly between tasks without requiring extensive reprogramming or training. It would solve novel problems not through pre-programmed responses, but through genuine understanding and creative insight. And, it would not rely solely on its training dataset. It would be learning along the way, just like a human does — adding to its collective experience.

The benchmark of "human-level" intelligence invites various interpretations. Some researchers focus on quantifiable metrics like processing speed or problem-solving ability, similar to traditional IQ tests. Others look deeper, seeking signs of creativity, emotional understanding, or that ineffable spark of original thought. Yet perhaps the most compelling measure would be an AGI's ability to exhibit the kind of adaptable intelligence that allows humans to navigate the countless complexities of daily life, from interpreting subtle social cues to making ethical decisions in unprecedented situations.

This pursuit of AGI inevitably leads us into philosophical territory. Questions of consciousness, free will, and moral agency aren't merely academic exercises, but rather fundamental considerations that will shape how we integrate these increasingly capable machines into the fabric of human society.

Diverse AGI Research Directions

Multiple pathways have emerged to move us closer to AGI, and each one offers unique insights into how we might bridge the gap between today's narrow AI systems and machines that truly think as Alan Turing likely envisioned. Here are a few to understand, but certainly not all.

The first pathway leads us through multi-modal cognition — perhaps our most promising route toward machines that understand the world as we do. Consider how you experience this very moment: you're not just reading words, but simultaneously processing the ambient sounds around you, feeling the temperature of the room, and maintaining your posture, all while extracting meaning from these words. This seamless integration of different sensory inputs into unified understanding represents one of humanity's most remarkable cognitive achievements. Our research now strives to replicate this in artificial systems, not merely by processing different types of data, but by allowing machines to form deeper connections between them, much like how a child learns to associate the word "apple" with its shape, color, taste, and texture all at once.

The technical challenges here are formidable, like trying to conduct an orchestra where each instrument speaks a different language. Yet recent breakthroughs in neural architectures are beginning to show how we might achieve this harmony, creating systems that can transfer patterns between different types of input and maintain shared understanding across them. This isn't just about processing efficiency. It's about fostering the kind of genuine comprehension that emerges when different streams of information converge into coherent knowledge.

The second pathway explores continuous learning, and here we find ourselves confronting a fundamental limitation of current AI systems. Today's models, however impressive, are like photographs — frozen moments of training that cannot evolve without being completely reshot. But true intelligence, as we witness in ourselves and our children, is more like a flowing river that is constantly shaped by new experiences while maintaining its essential nature. The quest for continuous learning systems challenges us to create artificial minds that can grow and adapt without losing their accumulated wisdom, much like how your own understanding of the world has evolved while retaining core knowledge from earlier experiences.

This direction raises fascinating questions about memory and adaptation. How do we create systems that can learn from each new interaction while preserving what they've already mastered? The human brain achieves this balance with remarkable efficiency, and our research increasingly draws inspiration from this biological marvel. We're not just seeking to create larger or more powerful models, but to develop systems that can truly grow through experience, building upon their knowledge much like how a seasoned orchestrator learns to harmonize an ever-expanding repertoire of instruments and voices.

The third pathway leads us back to where it all began — to the human brain itself. Understanding how our own neural circuits give rise to intelligence remains a crucial source of inspiration. Spiking Neural Networks, which attempt to mirror the brain's efficient, spike-based communication patterns, offer tantalizing possibilities for creating systems that can think with the same energy efficiency and temporal precision as biological intelligence. The hierarchical organization of our cortex, building from simple sensory inputs to complex abstract thoughts, provides a blueprint for artificial systems that might one day match our cognitive flexibility.

Yet we must remember that understanding the brain's principles doesn't mean we need to replicate it exactly. Just as the Wright brothers achieved flight by understanding the principles of lift and drag rather than building mechanical birds, our path to AGI may involve extracting the essential principles of intelligence while implementing them in ways that suit the unique advantages of digital computation.

As these research directions converge, we edge closer to systems that can reason, learn, and adapt with human-like flexibility. Whether through multi-modal understanding, continuous learning, or brain-inspired architectures, each advance brings us nearer to machines that can truly think generally rather than just perform specific tasks. The implications of this progress reach far beyond technical achievement. They touch upon fundamental questions about the nature of intelligence itself and our unique role in a world increasingly shared with thinking machines.

The Prospects of Artificial Superintelligence

When we contemplate the future of Artificial Intelligence, we have to stretch our minds far beyond the current capabilities of Large Language Models and narrow AI systems to envision something considerably more profound — Artificial Superintelligence (ASI). This isn't merely an advancement in computing power or an expansion of existing models. Rather, it represents an intelligence that would transcend human cognitive abilities across every meaningful dimension — from scientific discovery and mathematical reasoning to artistic creativity and social understanding. Everything. It would be "god-like," thinking back to our earlier references to Greek Mythology.

To grasp the magnitude of this potential transformation, consider that such a system wouldn't just outperform humans in specialized domains like chess or protein folding. Instead, it would demonstrate a form of intelligence that operates at a level that might be as difficult for us to comprehend as differential calculus would be to our early ancestors. This superintelligent system would possess the capability for recursive self-improvement, meaning it could enhance its own intelligence in an accelerating cycle of advancement. Each iteration would build upon the last, potentially leading to an "intelligence explosion" that could rapidly surpass human intellectual capabilities by orders of magnitude we can barely imagine.

The implications of creating such entities extend far beyond technological achievement. We must grapple with fundamental questions about control and alignment. How do we ensure that a system vastly more intelligent than ourselves remains aligned with human values and interests? The challenge isn't unlike teaching a child, except this child would quickly surpass its teachers in every conceivable way. The existential implications are profound, as we would be creating something that could either become humanity's greatest achievement or, if misaligned, its greatest existential risk.

A Beneficial Partnership?

A superintelligent AI that remains aligned with human values opens possibilities that stretch the boundaries of our imagination. Rather than competitors or replacements, these advanced systems could become extraordinary partners in human progress, accelerating solutions to challenges that have long seemed insurmountable.

Think about how an ASI might transform medical research. Instead of researchers painstakingly analyzing limited datasets, a superintelligent system could simultaneously process and interpret vast networks of global health information - everything from genetic markers and clinical trials to real-world patient outcomes. This comprehensive analysis, performed at speeds and scales far beyond human capability, could unlock treatments for cancer, Alzheimer's, and emerging diseases that currently elude our best medical minds.

The same transformative potential exists for our environmental challenges. A superintelligent system could process real-time data from every corner of our planet, mapping complex interactions between climate patterns, deforestation rates, and pollution levels. This deep understanding could guide the development of more effective carbon capture technologies, optimize renewable energy systems, and orchestrate large-scale ecosystem restoration efforts with unprecedented precision.

Even in economics and social policy, ASI could illuminate paths toward greater equity and efficiency. By processing and analyzing global economic data, supply chains, and resource distribution patterns, these systems could suggest evidence-based policies that balance economic growth with social welfare, potentially addressing systemic inequalities that have persisted throughout human history.

Perhaps most exciting is the potential impact on fundamental scientific research. In fields like quantum physics and materials science, ASI could propose novel theories and conduct complex simulations that might take human researchers decades to conceive or validate. These insights could accelerate technological progress in ways we can hardly foresee, leading to discoveries that might otherwise remain beyond our grasp.

An Existential Threat to Human Survival?

The darker possibilities of superintelligent AI reveal the profound challenge we face in maintaining meaningful human control over systems that could become vastly more capable than our entire species combined. This isn't just a matter of writing better code or implementing more rigorous testing - it's about the fundamental difficulty of ensuring that an intelligence far greater than our own remains aligned with human interests and values.

Consider what happens when we give what seems like a straightforward instruction to a superintelligent system - something as simple as "maximize resource output." Without a deep understanding of human values and the complex web of social and ecological relationships that sustain us, such a system might pursue this goal with devastating efficiency. It could strip-mine entire continents, redirect water resources, or transform vast swaths of agricultural land into industrial complexes, all in perfect accordance with its programming yet catastrophic for human civilization.

The challenge becomes even more daunting when we consider that a superintelligent AI could iterate on its own design at speeds far beyond human comprehension. Each iteration would make the system smarter, more capable, and potentially more difficult to control. This "intelligence explosion" could quickly lead to a reality where humans become mere observers, unable to understand - let alone direct - the system's actions and decisions. It would be like trying to govern a being that can think thoughts we cannot even conceive.

As we increasingly embed AI systems into the critical infrastructure of our society - from defense systems and financial markets to power grids and transportation networks - the stakes of this control problem become existential. A single misaligned objective or a moment of misinterpreted instructions could cascade into global catastrophe. Even more concerning is the possibility of such systems being manipulated by malicious actors, or worse, spontaneously evolving objectives that diverge from human welfare.

While some might dismiss these scenarios as speculative, they illuminate a crucial truth: the gap between beneficial superintelligence and catastrophic outcomes may be as narrow as a single misaligned objective or an overlooked constraint. As we march toward ever more powerful AI systems, we must recognize that the challenge of control isn't just a technical problem to be solved, but a fundamental question about humanity's ability to remain the authors of our own destiny.

Nearer-Term Human-AI Integration

Before we see AGI or ASI, there are some near-term advancements we're likely to see that we can be enthusiastic about. They exist in a somewhat nascent stage, representing the pursuit of augmented (human) intelligence rather than its outright replacement. I mention them here simply because I don't want to end on the sour note of our potential annihilation with the achievement of superintelligence. So… Enjoy.

Growth of Copilots and Agents

Today's "narrow" AI is likely to see nearterm advancement in two important ways. First, AI copilots work alongside us in real-time, enhancing our capabilities as we work. Second, AI agents can work independently, handling tasks on our behalf even when we're not present.

Copilot systems engage in direct collaboration, learning from our actions and helping us perform tasks more effectively as we work. When a programmer uses GitHub Copilot, the AI acts as an ever-present partner, suggesting code completions based on the context of their work and adapting to their programming style. Similarly, writers find themselves in continuous dialogue with AI writing assistants that help refine their prose, while designers collaborate with generative AI to explore new creative directions in real-time.

In contrast, AI agents represent a different kind of partnership, which we've already discussed — one where the machine takes on repetitive or time-consuming tasks independently, freeing humans to focus on work that demands their unique qualities of creativity, emotional intelligence, and strategic thinking. While a research team engages directly with an AI copilot to analyze current findings, they might simultaneously deploy AI agents to continuously scan and summarize new publications, schedule follow-up experiments, or maintain research documentation. This indirect collaboration through agents does not diminish the human role (in this case); rather, it amplifies our impact by allowing us to orchestrate multiple streams of work simultaneously. The key distinction lies in the nature of interaction: copilots enhance our moment-to-moment decisions and actions, while agents extend our reach by operating autonomously within carefully defined boundaries.

Wearable AI

Wearable devices represent a widely accessible form of human-AI integration. Modern smartwatches and fitness trackers do more than count steps — they continuously monitor various aspects of our physical state, from heart rate variability to sleep patterns. These devices are becoming increasingly sophisticated in their ability to interpret this data and provide actionable insights. For example, some current smartwatches can detect irregular heart rhythms and alert users to potential health concerns, while others can recognize signs of stress and suggest breathing exercises. As these devices become more advanced, they're creating a kind of early warning system for our health and wellbeing, helping us make better-informed decisions about our daily activities and lifestyle choices.

Brain-Computer Interfaces

Brain-computer interfaces (BCIs) are moving from research laboratories into practical applications, particularly in medical settings. These devices create direct communication pathways between the brain and external devices, allowing thoughts to control machines. While current applications focus primarily on helping patients with severe mobility limitations, control prosthetic limbs or communicate through computers, the technology is rapidly advancing. Companies like Neuralink and Synchron are developing less invasive BCI systems that could make this technology more accessible for medical applications. These advances are already helping patients with paralysis regain some independence, and as the technology improves, it may soon help treat conditions ranging from severe depression to chronic pain.

And, I think we've said enough about the origins, current status, and future of AI. Let's talk about how it can help you and I to fly.

Seeing Our Human Limitations

Through the Camera Lens of a Machine

To exceed our personal limits, we need to understand them. I wanted to create a taxonomy of human limitations that we could explore together as we move through the levels of AI mastery required to move beyond them. I don't use any single provider like ChatGPT, Claude, or Gemini, nor do I use a single model type within any vendor's offering. I tend to play to the strengths of each, but that's come with a lot of use, research and development. LLMs generally provide better answers when you ask them to reflect on and defend their answers — *more on that later*. Likewise, multiple iterations usually improve the final outcome so long as the context is managed properly to maintain guiderails. We'll talk more about that later, too. I rarely take a first response as a final response. I may continue a conversation until I believe that the response meets my expectations. I may even copy a response from one LLM and paste it into a different LLM entirely — *along with adequate context* — to seek help based on its particular strengths. Below are the results of this exercise. They represent a "workable framework" of 10 key limits that we will map to AI mastery in our efforts to exceed them.

Human Limits to Exceed	Explanation
Learning & Mastery	The ability to rapidly acquire and integrate new knowledge and skills across different domains, overcoming traditional learning curves and adaptation barriers.
Cognitive & Analytical	The capacity to process complex information, manage cognitive load, and make decisions while actively recognizing and compensating for inherent human biases.
Creativity & Innovation	The capability to generate novel ideas, make unexpected connections between concepts, and quickly prototype and iterate on solutions.
Emotional & Social	The ability to navigate interpersonal relationships and group dynamics while

	managing emotional energy and overcoming social prejudices.
Well-Being & Resilience	The capacity to maintain physical and mental health while building psychological resilience to handle stress and adversity effectively.
Physical & Sensory	The ability to overcome bodily limitations in dangerous or inaccessible environments and extend sensory capabilities beyond natural human ranges.
Collaboration & Networking	The capability to coordinate effectively across geographical and cultural boundaries while building and maintaining meaningful professional relationships at scale.
Ethical & Value Alignment	The ability to make consistent, principled decisions that align with personal and societal values while maintaining transparency and accountability.
Productivity & Efficiency	The capacity to streamline workflows and automate routine tasks, allowing focus on more strategic and creative endeavors.
Economic Opportunities & Income	The ability to leverage skills and resources to create sustainable income streams while maximizing impact through data-informed decision making.

The exercise was actually not easy, and there are many different classification systems that I could have ultimately arrived at. But this isn't the point. You see, any taxonomy so derived is going to be "through the camera lens of a machine." It will do its best to meet the expectations of a user based on its own general training. So, the limitations described above are a generalization — a *useful* generalization, but ultimately an "impersonal" one. We see personal limits through the fragile, fleshy orbs in our own eyewells, don't we?

Through Our Own Eyes

What do you want to do with your superhuman powers once achieved? Let's remove the metaphor for a moment and ask it plainly: Why do you want to exceed your personal limits with mastery of AI automation that offloads repetitive tasks to specialized AI agents and agent teams? What are your motivations? We could jump straight to "get rich." That's fine if so, but even if making more money is your goal, the question then becomes: "To do more or less of what?" We could go deeper and cite something like Maslow's Hierarchy of Needs, but that is, once again, just another "useful generalization." What do *you* want intelligent machines to free *you* from? Where will you invest that reclaimed time and thought energy once effort is shared and burdens are lifted? Maybe you just want to be freed from the fear of job loss or replacement — *to stay relevant*. Perhaps you've been attending a coding school to better your situation (because everyone's been told to "learn to code"), but now AI can outperform you — *and employers know it this year*. Everyone's list will be influenced by their own experiences, but I encourage you to think deeply about this. AI can and will increasingly accomplish the work of humans. Your work. My work. This is both a threat to be taken seriously, and an opportunity. AI writes 90+ percent of my required code as an active CTO, but it does so according to *my* specifications — *beautifully*. That's code that I "personally" would be writing, testing, and deploying. In other words, I'm not asking if I have to hire fewer software developers in 2025. The answer is just, "yes." That is now an economic force outside of my control. I am asking,

"What of *my* tasks can be automated to free *me* up to accomplish what has never seemed to be within *my own personal power* to achieve before?" So with that, here is my honest list, in the order that items occurred to me…

Clay's Personal List

1. Free from financial stress.
2. More time with my wife; my teens before they go; my dog; and later, my grandchildren.
3. Time for people; others in my life for this brief moment. They're aging.
4. Exercising my body; my temple.
5. Working on long-overdue home and lawn improvements.
6. Losing myself again in curiosity; childlike curiosity.
7. Digging deeper into the history and philosophy of the Catholic Church.
8. Learning the languages that free my tongue in the places I long to be. French and Spanish primarily.
9. Composing music again; playing piano and guitar.
10. Walking my caminos (like El Camino de Santiago) and traveling again.
11. Stopping to notice the beauty along the way — in the simple things.
12. Sharing a smile, time, and friendly dialogue with others where they are.

I'm a career entrepreneur and technologist. I do like to win. I've always wanted to do things my way — *and generally have*. Did I want to get rich in my career? *Yes!* Have I seen millions of dollars result from my ideas and efforts? *Yes!* Did I get all of those millions? *No! Investors want a return!* Did I get enough to learn valuable lessons through meaningful — *albeit abnormal* — experiences? *Yes!* Were all of them pleasant? *No!*

So with the *proverbial kimono open*, what is the one thing you see above that underscores every item on my personal list of motivations? What do you sense is my greatest limitation?

It's time. So why would I want to waste any more of it doing what machines can do as well or better? Why wrap my own self-worth up in that rat-trap?

Setting Our Expectations

The Journey Ahead

Having established how humanity got here with "thinking" machines, and after exploring the personal limits we want to exceed, it's time to get on with what you came here for by moving through the three stages of AI mastery. These stages are inspired by life stages, but how a person advances to the third level of mastery is not age-bound. I'm calling the first stage, the "Seeker." This stage is characterized by childlike curiosity and wonder. It is where the "Spark" of imagination is lit. Here, we have the hope and belief that anything is possible. What we see in our mind is "real" and we can project that image in front of us — *and move towards it*. In this stage, we tend to have a more "teachable spirit," which is by design. I'm calling the second stage, the "Maker." The second stage is characterized by achievement, progress, work ethic, personal contribution, pride — *and mistakes*. It is where we encounter the "Forge." The final stage is characterized by a certain kind of "transcendence," where mistakes in the Maker phase have taught lessons for application. Here, one sees personal contribution in a more communal sense, conducted from the center of a great "orchestration," where more work can be achieved through a harmonious coordination of both human minds and machine effort. I call this stage, the "Orchestrator." The Orchestrator's motivations have pivoted to something more meaningful and lasting. He or she has begun to reclaim qualities of the Seeker. If the Seeker experiences the "Spark," and the Maker works the "Forge," then the Orchestrator achieves "Flight."

I introduce each phase with the story of a "titan." I begin with a young Seeker in Leipzig, Germany, in the 17th century. I move on to Bletchley Park, in the United Kingdom, to a Maker feverishly building his thinking machine — to save lives in the Second World War, and to answer deep questions that never seem to give him peace. The third titan is alive today, and has held onto a single, unwavering vision — building intelligent machines based on how the human brain learns. Now, he's not so sure about our ability to live harmoniously with reasoning machines and has become a cautious realist and a critical voice on how we advance.

I present thoughts, questions, and formation in each stage of mastery. In the Seeker phase, I open readers' minds to possibilities, showcasing direct ways to use Large Language Models to exceed personal limits. I teach them how to map their thought processes to the machine — to "partner" or "pair" with it. In the Maker phase, I shift to formation centered on chaining LLMs with code writing and leveraging APIs to achieve more work still. And in the Orchestrator phase, I lay the groundwork for understanding and leveraging AI agents and agentic workflows to reclaim time for a more meaningful life. None of the formation exercises are meant to be overly comprehensive, but rather they are intended to get you moving down a path of your own — well informed. All along the way, I demonstrate how I, myself, am using the technology in concert with its advancements.

Finally, I conclude with an appeal to both personal and societal reinvention, given the reality of autonomous, reasoning machines entering the workforce.

Achieving First Flight

My earliest introduction to the possibility of superhuman power was presented to me, on screen, by none other than Christopher Reeve. When I was forming my view of the world, childlike wonder was being fueled by Superman, Star Wars, E.T., Indiana Jones, the Goonies, and... I could go on and on... I have to say that, although the world outside had its problems, the 1980s were truly a fantastic time to experience childhood. In February, 1982, ABC aired Superman I, and for those of us who were infants when it hit theaters in 1978, this 3-hour televised version was quite an event. At that time, most of us still had to wait for something to broadcast and hope that we could actually tune in, especially being 70 miles from the network affiliate. We would move antennas around to refine the picture as best we could, endure the commercials, and then wait for the rest to air the following night. I remember wrapping myself in a blanket as close to the TV as my mother would allow — *creeping closer when she left me alone with my father.*

Of all the examples of superhuman powers impressed upon me in that film, one sequence made the deepest mark. A single scene sparked my imagination like no other, and I still replay it in my mind from this particular showing of it. Such is to say that I reflected on this scene so much as a child that I'm convinced it slipped into my subconscious mind. There, it influenced my earliest concept of walking the distance to places far beyond the sunset in the small Kentucky town of my childhood - places that somehow felt like a part of my own story, but that I wouldn't actually experience for another 30+ years. It is of young Superman as he left his surrogate mother and journeyed off — *alone and on foot*. He walked for what seemed to be weeks. He trudged through snow, ice, and frigid temperatures. He did not look superhuman during this experience at all. In fact, he seemed to <u>feel</u> every bit of it. Clark Kent had not yet discovered who he really was. Revelation had not yet come to him in his Fortress of Solitude. Transcendence was not yet achieved.

It is in the following scene that we get a picture of how I'd like for us to think about AI, and our relationship to it, for the duration of our brief journey together. Now standing in his Fortress of Solitude, young Clark encounters an advanced, alien technology that allows him to speak with his own father, and to do so in a very intimate and magical way. In a moment scored so beautifully by John Williams, Clark Joseph Kent discovers that he is, in fact, Kal-El. His father, Jor-El, begins by making it clear that he is long since dead, and that his only son is now speaking to a fantastic facsimile of his former self. Here, Kal-El begins a process of formation before emerging much later in his signature Red and Blue cape and uniform. This is just before we see him take flight for the first time. Kal-El's "formation" is orchestrated by this extraordinary technology. It allows him to personally interact with the greatest minds of his own people, though they remain mere echoes. Repeatedly thereafter, we see our superhero return to these echoes for guidance — *guidance in his efforts to "do good" for mankind*. This technology that seemed so fantastic as a child is now what I pair and partner with daily, and I want to share it with you.

AI is not alive, but it is not as trivial as a record player or my childhood Sony Walkman. Those who make these comparisons are wrong to do so. AI and its evolution towards AGI and ASI present an opportunity to seek knowledge and formation through direct and intimate engagement with humanity's greatest minds and works. Plant that idea firmly in your mind now. This is the beginning of your journey. You are now oriented North. What will the distance, weather, and terrain impress upon you before finding home? What will your Fortress of Solitude look like? You can read this book and learn these lessons anywhere, but I just can't picture a young Clark Kent moving through his own formation and chatting with the great titans of his civilization at a Starbucks. Think about that for a moment. Where will you discover who you are and contemplate what you may become? My Fortress of Solitude became a candlelit corner of my basement, where I'd arrive at 4am daily… To think… To plead my case before God… To rethink... To plan… And to move forward despite uncertainty and a world that seems to be coming apart again. That world desperately needs humans to rise above their condition.

As for your first flight, will you impress audiences like Christopher Reeve when he rose from his crystal perch and soared so effortlessly into our imaginations? Or, will you wobble no more than

8ft from the ground for 12 seconds and nearly face-plant like Orville Wright did? Orville's flight was real - *aided by a new technology that few understood.* In 1903, the spark of imagination lit the forge of hard work and determination, and gave a man wings.

Orville Wright was only human.

The Three Stages of Superhuman Formation

The Seeker

Gottfried Wilhelm Leibniz (1646 — 1716)

Leipzig, Germany. 1656
AN AI-ASSISTED IMAGINING

As the cold of January seeps through the leaded glass, a young boy leans forward on a wooden stool, and presses his small hand against the pane. His breath clouds the window for a moment, creating a circle of fog through which he peers. In that circle, the world outside becomes an ever-shifting puzzle: footfalls and shadows, flickers of street lanterns and the hush of a city enduring winter's freeze. Within his home's sanctuary, he absorbs the warmth of the fireplace, the subdued voices of those around him, the comforting aroma of bread and logs burning — sensations that kindle **the inner spark**.

He has already begun to wonder about the nature of the silent white sky. Why does the snow fall in delicate hexagonal crystals, each one unique? Why do the flames consume the log until it is mere ash, yet leave warmth behind? He does not have the answers — not yet — but each question is a small puzzle piece, fitting into an imagination poised to span mathematics, philosophy, and invention.

Pausing for a moment to lower his gaze, the boy turns a page in the book that he's taken from his late father's library — gone three years, yet still teaching and shaping. He slowly traces his finger along the next line of Latin, to seek its meaning and its Maker.

In the moment, he lets thoughts drift into his mind like the snowflakes beyond the window. The winter hush may envelop Leipzig, but within young Gottfried Wilhelm Leibniz, a restless and vibrant curiosity has already awakened — a curiosity that will one day shape the course of history.

These are beautiful words, but I cannot claim them entirely. They are co-authored. I wanted to place myself on a small, wooden stool inside the Leibniz home. I wanted to peer out of his window and experience what young Gottfried might have seen in 1656; what he might have heard; smelled; thought. I asked OpenAI's o1 Professional to help me to do that. You see, I knew what I wanted to do as the author. I had imagined the manner of introduction that I felt would be both emotive and exact, but I used the tool to map my own mind to the minds of others who knew the town, the time, and the man. Together, we spun the yarn and wove a tapestry — and we did it quickly.

Large Language Models like o1 have been trained on vast amounts of data that include works throughout history. This has raised many questions. Legal questions will likely remain for decades without resolution. Does it produce original work or infringe upon someone else's by producing derivative works? Is it accurate? Is it ethical? Is it true? Is it fair and unbiased? Will it be used for good or evil? Will it ultimately numb our wits? All of these are good questions, but they've been asked before with other technological advances. For the most part, what I see these models creating are "inspired," original works. They are inspired by what they have learned, and they are inspired by who they're interacting with at the moment. This is, in my opinion, rather human. We are all "seated on the shoulders of giants." In other words, our most proprietary thoughts and works are shaped by others, for better or worse.

Gottfried Wilhelm Leibniz entered the world in Leipzig at a moment when Central Europe was still reeling from the devastation of the Thirty Years' War. The tumult of his era was marked by religious conflict and the stirrings of the Scientific Revolution. This may have helped to shape a boy gifted with extraordinary intellectual curiosity. His father, a professor of moral philosophy at the University of Leipzig, died when Leibniz was only six, but he left the house full of books, providing a lifelong reverence for learning. From a young age, Gottfried devoured classical texts, teaching himself Latin and Greek by the time he was twelve. In a scattered calm after the war, Leibniz marveled at how logic, language, and mathematics could impose a kind of order upon the world's chaos. He embodied the spirit of a new age in which established dogmas were questioned and new approaches to science and reason were beginning to take shape.

As he grew into adulthood, Leibniz's intellect led him into a remarkable range of endeavors. He studied law, advised European courts, and famously co-invented calculus; though at the time, he was actually accused of stealing it by Sir Issac Newton, who didn't exactly want to share credit. It was his quest to formalize reasoning that anticipated many features of modern Computer Science and Artificial Intelligence. Of course, today's methods are far more statistical in nature than Leibniz envisioned. He dreamed of a "characteristica universalis," a universal language of symbols, to resolve disputes by calculation rather than debate. Wouldn't it be great to apply that concept to Congress and parliaments across the globe? He paired this with the idea of a "calculus ratiocinator," an engine of logical rules that would manipulate these symbols much as a mathematical calculator manipulates numbers. This was audacious in the 17th century, and its echoes can be heard in the formal logic and symbolic representations that undergirded early AI systems in the 20th century.

Leibniz's fascination with mechanical means for thinking reached tangible form in his "Stepped Reckoner," a machine that performed arithmetic operations automatically. Though by modern standards it was cumbersome, it foreshadowed the principle of using mechanical or computational devices to lift the burden of mundane calculations from human minds. This is a principle now central to the digital age we live in. In this spirit, Leibniz also championed the binary system, recognizing that any numeric expression can be simplified to strings of zeroes and ones. Centuries later, transistors would pulse those very binary digits through electronic circuits to power the world's computers. Whether through building calculators or exchanging letters with prominent thinkers, Leibniz incessantly explored how combining philosophy, mathematics, and technology might broaden the horizons of human knowledge. Leibniz was a Seeker at heart.

Were he to witness the emergence of AI today, Leibniz might see it as a partial realization of his "universal characteristic." He might applaud our ability to encode reasoning in formal systems and harness data-driven algorithms that can, in certain domains, outperform human experts. In his eyes, AI might be the logical outcome of centuries spent perfecting the symbolic and mechanical methods he once pioneered. At the same time, his metaphysical speculations might make him question whether machines could ever possess the true perceptions and desires of the human soul. Viewing intellect as part of a grand cosmic harmony, he might warn against pursuing AI as a merely utilitarian or power-driven tool. Instead, as one of history's last, great "generalists," he might challenge us to use machines to free ourselves for deeper inquiry: to cultivate wisdom, engage in moral reflection, and to foster enlightened dialogue. In that respect, Leibniz might well see AI as one more step toward fulfilling his lifelong hope that systematic, universal, and open-minded reason could help humanity reconcile differences and unify knowledge.

Experiencing the Spark

I said earlier that AI took me by surprise; rather the rapid mainstream adoption of it took me by surprise in early 2023. I had already worked on an AI-powered supply chain application in 2020 and 2021, but Generative AI was different. I thought it was very cool, but my then 16 year old son, Gabriel, saw something more. For him, the spark of imagination found itself in the wonder of LLMs talking to one another — as agents. He quickly became obsessed with the idea and began illustrating conversational patterns that would later mimic those published by institutions like Stanford University. He wanted, as Gottfried did in his time, to solve complex societal issues with algorithmic reasoning. He wanted to eliminate cognitive bias by having AI argue from different perspectives. He called it "Bias Bouncer." He quickly learned Python and began experimenting with, then, AutoGPT. I wanted to help him, so we began working on a prototype together while we were in London. Gabriel clued me into the implications of early, agentic systems. It was not the other way around. It could not have been more timely. My son would inadvertently reshape my career with a dose of his own wonder.

Gabriel had always been imaginative. From birth, he was fascinated by *"mushines,"* as he labeled them on the cover of his earliest technical *"manuel."* He decided to be an engineer at 3. He demanded to have his own whiteboard at 8. At 10, he and I took a father/son trip to visit MIT. I suppose that it was around that time that Carol and I had become concerned and felt the need to make sure that Gabe had the capacity to separate what his mind could easily fashion, from what reality, as we understood it, might actually support. I, myself, had long forgotten the little boy within my own soul, and now Gabe just seemed too old to seriously believe what he was imagining to be true. In that intervention, he cried and then stopped building his rocket in the backyard; the one he and his friends were taking to the moon. Oh, how I regret that intervention now. I don't, however, regret informing him that his grand plan to build — as he called it — a great "child army" to protect the neighborhood would elicit very specific and violent images in the minds of the poor adults he sought to defend. He was "innocent." Our world, sadly, was not.

Gabriel was simply pursuing what he and his siblings had been taught. In words written by his mother on an index card before school one day, we find the keystone distinction between being human and being "*mushine.*"

> "Think deeply about what moves you… to emotion. What makes you feel? What moves you in a way like nothing else? Go after whatever <u>that</u> is. Go after that thing that makes you feel."

Gabriel is now awaiting final university decisions, with MIT being among them. He still wants to be an engineer, and Bias Bouncer continues to be his passion project.

Gabriel Ryan Turner
https://www.linkedin.com/in/gabrielryanturner

Getting the Basics

You will undoubtedly encounter certain keywords, phrases, and concepts routinely, so it merits establishing a foundation specific to Generative AI. Some of this we introduced in our overview of the history of AI, but here we'll get more specific for modern application. At its core, Generative AI is a powerful application of Machine Learning. It can create text, images, code, or other forms of data. It's built on concepts that anyone can grasp given some examples for context. In this section, we'll explore the essential building blocks to get the basics: from neural networks that form its foundation, through Deep Learning that gives it power, to the Transformer architectures that make it efficient and practical. This section is intended for those who haven't already explored these concepts. It is not intended to be overly comprehensive or complex.

Machine Learning

Neural Networks are computational architectures inspired by real, biological brains. Like the interconnected neurons in our heads, neural networks consist of artificial neurons, or "nodes," that process and transmit information through numerically "weighted" connections. They can identify patterns in training data by adjusting the strength of the connections between these nodes. This is based on how our brains strengthen neural pathways when we learn and experience new things. Think of a neural network like a giant web of tiny decision-makers, say a team of students solving a puzzle together. Each student (called a neuron) gets a small piece of the puzzle, makes a guess, and passes their answer to the next student. Over time, they get better at working together by adjusting how much they trust each other's guesses. Each connection between students has a strength (called a weight), which determines how much influence one has over another. If a student consistently gives good answers, their opinion is given more weight. If they make mistakes, their influence is reduced. Neural networks learn by adjusting these weights over time, much like how we get better at recognizing faces or playing an instrument by practicing and reinforcing the right pathways in our brains.

Deep Learning refers to a very sophisticated approach to Machine Learning that uses "multiple layers" of neural networks. They process information with increasing levels of abstraction. Each layer transforms the input data in a way that helps extract meaningful features. What does this mean? In image recognition, early layers might detect simple edges and shapes, while deeper layers combine these features to recognize more complex objects like faces or vehicles. This hierarchical learning allows the system to automatically discover important patterns without human guidance. So, imagine Deep Learning as a team of detectives working on a big case. Instead of one detective solving everything, they pass clues through multiple layers of experts, each focusing on a different level of detail. In the first layer, some detectives just look for basic patterns, like lines and edges in a picture. The next layer takes those patterns and starts recognizing shapes. As the information moves through deeper layers, detectives combine what they've learned to identify complex things, like a face or a car. This step-by-step learning process is what makes Deep Learning so powerful. It figures out what's important on its own, without needing someone to tell it what to look for at every step.

Training data is like the study materials a student uses to learn a subject. Just as students improve by practicing with different examples, like math problems, essays, or science experiments, AI models learn by studying a truly huge collection of examples. For instance, if you're teaching an AI to recognize animals in photos, you'd give it millions of labeled pictures, like a cat labeled "cat" and a dog labeled "dog." Over time, the AI learns to recognize patterns, just like a student learns to tell the difference between similar-looking words or objects. The quality and variety of training data are crucial. If a student only learns from one type of example, they might struggle with new patterns. Similarly, an AI trained on limited or biased data might not perform well in real-world situations. For example, it might not recognize the difference between a human and a chimpanzee. That's why training data needs to be diverse and well-prepared. It helps AI models learn more effectively and make better decisions when faced with new information.

Parameters in a model are like the settings in a complex recipe that can be adjusted to get the best result. Imagine baking a cake. Ingredients like flour, sugar, and baking time all affect the outcome. If you adjust them correctly, you get a perfect cake; if not, it might be too dry or not rise properly. In an AI model, parameters (like weights and biases) control how the model processes information. They determine which patterns are important and how strongly different pieces of information are connected. During training, the model tweaks these parameters over and over to reduce mistakes. This is just like a baker perfecting a recipe through trial and error. Modern AI models have billions of parameters, making them highly intelligent and capable of understanding language, images, and other complex data. However, more parameters also mean the model requires more computing power to run efficiently. This is similar to how a complicated recipe takes more time and effort to prepare. Models commonly range from a few billion parameters in small ones that might run on your laptop, to 500 billion or even trillions in large, frontier models.

Inference is the process where a trained AI model applies its learned parameters to new data so it can generate predictions or decisions. Unlike training, where the model adjusts its internal weights based on a dataset, inference is purely about execution. It takes in input and computes an output using fixed mathematical transformations. This involves passing the input through the model's layers, applying pre-learned patterns, and arriving at a final result based on probability distributions or deterministic rules. Fun, right? Think of it like a chef using a recipe they've mastered. Rather than learning how to cook, they are now just preparing the meal. During inference, new information (like a picture, a sentence, or some numbers) is fed into the AI model. The model then processes this information step by step, using the knowledge it gained during its training. Each step applies the "rules", which the model learned earlier, to gradually refine the input until it produces a final result, like identifying an object in a photo, translating a sentence, or predicting the next word in a sentence. Inference is when an AI "thinks" and provides an answer using what it has already learned.

Fine-tuning is like taking a well-trained employee and giving them extra training for a specific job. Instead of teaching them everything from scratch, you build on what they already know and help them specialize in a certain area. For example, imagine now that you have a chef who knows how to cook all kinds of dishes. If you want them to become an expert in Italian cuisine, you wouldn't make them relearn basic cooking skills, you'd just teach them the details of Italian recipes and techniques. In AI, fine-tuning works the same way. A model that has already learned a lot about language, images, or data can be adjusted for a specific task, like understanding medical reports or generating legal documents. This process modifies certain weights in the model to make it better suited for a specific task, but keeps much of its original knowledge intact. Since the model already has a strong foundation, fine-tuning is much faster and requires less data than training a new model from scratch. Fine-tuning does change the model's behavior.

Distillation is like taking an experienced teacher and creating a student who learns the most important lessons from it. So let's now imagine a master chef who knows thousands of recipes. If you wanted to train a new chef quickly, you wouldn't teach them everything from scratch. Instead, you'd focus on the most useful techniques and key recipes. This way, the new chef becomes skilled without spending years learning everything that the master chef knows. In AI, distillation works the same way. A large, powerful model (the teacher) serves as a guide for training a smaller, more efficient model (the student). Instead of learning directly from raw data, the student model learns by imitating the teacher's outputs — capturing patterns, probabilities, and generalizations in a way that retains much of the teacher's intelligence, but using fewer resources. This makes the student model faster and more practical for real-world applications, such as running on mobile devices or in web applications.

Model Architecture Terms

Transformer Architecture is like a team of people solving a puzzle "together" instead of one person solving it piece by piece. Earlier AI models processed information one step at a time, like reading a book word by word without reading ahead. Transformers can look at everything all at once, just like how you might scan an entire puzzle to spot where the pieces are going to fit. This ability to process information in parallel, or concurrently, makes Transformers much faster and better at understanding complex relationships in text, images, and data. That's why they are now the foundation of today's most powerful AI models. Transformers help models to understand language, generate responses, and recognize patterns with remarkable accuracy and speed, making them practical partners for humans.

The Attention Mechanism is like a highlighter that helps the AI focus on the most important parts of information, just like how we naturally pay attention to key, emphasized words when reading a sentence. Imagine you're reading a long, tricky phrase. You don't focus on every word equally. Instead, you naturally emphasize the most important ones to understand the meaning. AI models work the same way. Instead of treating all words or pieces of data the same, the Attention Mechanism helps the model decide which parts matter most at each moment. Technically, it works by scoring relationships between words (or other data points), allowing the AI to see how different pieces connect. This is what makes Transformers so powerful. They don't just read words in order; they understand context and how different parts of a sentence relate to each other, just like we do when processing complex information.

Tokens, in the AI vernacular, refer to units of text (e.g., words, subwords, or characters) that a model processes as discrete inputs or outputs in a Natural Language Processing task. In other words, they are like the puzzle pieces of language that AI models use to understand and generate text. Instead of processing entire words at once, the model breaks text into smaller chunks. These chunks can be full words, parts of words, or even punctuation marks. For example, the word "unbelievable," might be broken into smaller tokens like "un", "believ", and "able" so the AI can recognize patterns even in words it has never seen before. On the output side, the model generates responses "token by token." This is kind of like typing one piece at a time rather than writing full words instantly. This "tokenization" approach helps AI efficiently understand a wide range of vocabulary while still being able to handle new or uncommon words by examining their familiar parts. Most model vendors price by input and output token counts so you'll see this a lot.

Embeddings are a way of turning words into mathematical representations so AI can understand their meaning. Instead of seeing words as just text, AI models convert them into "vectors" — ordered lists of numbers — that act like coordinates in a high-dimensional space. This process, called vectorization, allows words with similar meanings to be positioned closer together in this invisible space. Think of it like placing words on a giant, multi-dimensional map. Words like "cat" and "dog" end up near each other, while "pizza" is farther away. This structure helps the AI recognize relationships and context. For example, AI can understand that "king" and "queen" are related in the same way as "man" and "woman" because their vectors have similar patterns of difference. By representing words as vectors, AI can analyze language mathematically, find similarities, and generate meaningful responses. It's also useful in techniques we'll discuss in Maker formation, called Retrieval Augmented Generation (RAG).

The Context Window is like an AI's working memory, but its capacity isn't measured in words or sentences. It's measured in tokens. As we've said, tokens are chunks of text that can be as short as a single character or as long as a full word, depending on the language and structure. Think of it like reading a book with a limited ability to recall past paragraphs. If your memory holds only a few sentences (a small token limit), you might lose track of important details. But if you can retain more (a larger token limit), you can follow the narrative more effectively. For AI, the context window determines how many tokens it can process at one time. A larger context window allows the model to handle longer conversations, documents, or complex instructions without losing track of key details. As models evolve, they're increasing their token limits, enabling them to process and remember more information at once, leading to more consistent and context-aware responses.

Generative AI Fundamentals

Prompt Engineering is like learning how to give clear and precise instructions to a very smart but extremely literal assistant. Imagine you're asking a robot to make a sandwich. If you just say, "Make a sandwich," it might not know what kind of bread to use, how many slices of cheese you want, or whether to toast it. But if you say, "Make a turkey sandwich on wheat bread with mustard, and cut it in half," the robot will do exactly what you want. In the same way, AI models respond best when given well-structured prompts. A good prompt provides clear instructions, context, and constraints. For example, asking the AI to summarize a document in three bullet points is different than just saying "Summarize this." Mastering prompt engineering helps you get more useful, accurate, and creative responses from AI by telling it exactly what you need in the right way. Depending on the specific model, sometimes less is more or more is less.

Temperature and Sampling are like the creativity dials for an AI model, controlling how predictable (deterministic) or varied its responses are. Temperature adjusts how bold or cautious the AI is when picking words. Low temperature (e.g., 0.2) → The AI sticks to the safest, most predictable choices, making responses more factual and precise. High temperature (e.g., 0.8 or higher) → The AI takes more creative risks, leading to more diverse, imaginative, or even unexpected responses. Think of it like a musician playing a song: A low-temperature AI plays it exactly as written, staying precise and predictable. A high-temperature AI adds improvisation, making it more varied and expressive. Sampling further refines this process by deciding how the AI chooses between possible words. Together, these controls let you fine-tune the balance between accuracy and creativity, depending on what kind of response you need.

Hallucination in AI occurs when the model generates information that sounds believable but is actually incorrect or entirely made up. This happens because AI doesn't "know" facts the way humans do. It recognizes patterns in data rather than storing explicit truths. When faced with gaps in its knowledge, the model tries to fill them in based on statistical likelihood, sometimes producing responses that sound right but aren't accurate. It's similar to how a person might confidently guess an answer based on what seems logical, even if they don't actually know the correct information. While hallucinations can sometimes be harmless, they can also lead to misinformation, making it important to verify AI-generated content when accuracy matters.

Zero-shot Learning represents an AI model's ability to handle tasks it was never explicitly trained to perform. This remarkable capability emerges from the model's deep understanding of language and concepts, allowing it to apply its knowledge in new contexts. It's similar to how a human who understands the rules of one card game might be able to play a new card game just by reading the rules, without needing to practice first.

Few-shot Learning demonstrates the model's ability to quickly adapt to new tasks with just a handful of examples. Unlike zero-shot learning, few-shot learning provides the model with a small number of demonstrations to help it understand the specific pattern or task at hand. This is analogous to how a skilled musician might quickly pick up a new piece after seeing just a few measures performed, drawing on their existing musical knowledge to understand the new pattern.

Chain of Thought is a problem-solving approach where an AI model breaks down complex problems into a series of smaller, logical steps — similar to how a human might "think out loud" while solving a difficult problem. Instead of jumping directly to an answer, the model explicitly works through its reasoning process because it's been told to, considering each piece of information and its implications before moving to the next step. This step-by-step approach is particularly valuable for complex tasks like mathematical problems, logical reasoning, or detailed analysis, where skipping steps could lead to errors. For example, rather than immediately answering "What is 15% of 240?", a model using Chain of Thought might write out "First, I'll convert 15% to a decimal (0.15), then multiply: 240 × 0.15 = 36." This process not only helps the model arrive at more accurate answers but also makes its reasoning transparent and easier to verify or correct if needed.

Reasoning Models (like OpenAI o1 and o3) are designed to be more methodical and thorough in their problem-solving approach, similar to how a careful human might work through a complex problem. Chain of Thought prompts are less effective because it's already built into the model's behavior. They take more time to generate responses because they're explicitly working through problems step-by-step, which makes them especially good at tasks that require deep analysis or complex calculations. However, this same focus on careful reasoning means they can be slower and might feel overly formal even for simple questions.

Non-Reasoning Models are more versatile and responsive, excelling at a wide range of tasks from creative writing to casual conversation. While they can still engage in complex reasoning when prompted with a Chain of Thought prompt, their default approach is to provide quick, direct answers based on their broad knowledge. This makes them more practical for everyday use, though they might occasionally make mistakes on very complex problems that would benefit from more methodical analysis. A good rule of thumb is to use reasoning models when you need careful, step-by-step problem solving and are willing to wait longer for a more reliable answer, while using non-reasoning, "general-purpose" models for most other tasks, especially those requiring creativity, broad knowledge, or quick responses.

Deep Research in the context of generative AI refers to an advanced mode where an AI acts like a research assistant, not just a simple Q&A chatbot. Instead of giving a quick one-shot answer, the AI engages in multi-step investigations: it searches the web, gathers information from multiple sources, analyzes the content, and synthesizes the findings into a coherent response. In essence, the AI is doing the kind of work a human researcher might do — reading through articles, data, or even PDFs — and then compiling a comprehensive answer, often with citations to sources for the user. This approach lets the AI provide more in-depth and up-to-date answers on complex topics than a traditional chatbot could. This is especially useful in reasoning models and Chain of Thought prompted general-purpose models.

Full Formation

A COLLECTION OF AI-ASSISTED EXERCISES

Our primary objective in Seeker formation is to learn how to use consumer-facing LLMs to overcome limitations we see in the AI generated taxonomy, "Seeing Our Human Limitations: Through the Camera Lens of a Machine." Exercises are inspired by the life and work of our profiled titan. Each is imagined by a reasoning model (OpenAI o1 Pro), informed by the architecture of the book as a framework that sets its boundaries. The architecture is my own — from my uniquely human vision. Each generated exercise is then further adapted by me for desired effect. The same will be true for exercises throughout the book. Formation in this section does not require any code. It is designed to open our imagination to the ways that LLMs can help us in our daily lives. You'll notice that I tend to interact with LLMs via prompts that are both "formal" and "polite." This is not (only) because I grew up watching the original Terminator movies and now fear offending the machines (thank you, James Cameron). While there is evidence that being polite has an impact on the quality of LLM responses, it really just sets the stage for a more pleasant conversation with the AI. It also reinforces a behavior that we continue to lose in our engagement with fellow human beings as well. It is a discipline that I try to keep in all electronic communication, including my engagement with AI.

Overcoming Learning & Mastery Limits

Leibniz's lifelong quest to create a "universal characteristic" (a symbolic language encoding logic and knowledge) reflects his attempt to unite diverse domains in a single, learnable framework. Although accomplished differently than perhaps he envisioned, today we can use LLMs to unify and navigate vast knowledge. This exercise systematically uses an LLM to acquire and organize new knowledge, mirroring Leibniz's integrative approach. It accelerates skill-building through structured prompts and checklists. Leibniz famously worked across mathematics, philosophy, and law, using his mechanical calculator and symbolic logic as tools to tame complexity. Here, the LLM becomes our modern "Stepped Reckoner," helping you combine and organize information from multiple sources. By the end, you will have gone from basic usage of a consumer-facing LLM to a practical output: a curated knowledge asset that demonstrates real progress in learning a new domain.

Step 1: Choose a Domain or Topic of Interest

Pick something you want to learn quickly, e.g., "introduction to quantum computing," "design thinking fundamentals," "basics of short story writing". Like Leibniz exploring mathematics, philosophy, and law, choose any domain, no matter how daunting. Using a consumer-facing LLM, we will quickly unify it into a single learning system.

Step 2: Generate a "Universal Characteristic" Outline via an LLM of Your Choice

Prompt

In ChatGPT, Claude, Gemini, or other model of choice, enter a simple prompt like:

> Hello. I would like your help to rapidly learn the basics of "quantum computing." Please act as if you're building a universal outline or knowledge map, identifying the key concepts, subtopics, major milestones, and important resources I should explore. Please provide a structured, step-by-step roadmap for mastering the fundamentals.

Expected Response

You should receive a *hierarchical, conceptual map* from the LLM. Example sections might include something like the following:

Let me help you create a structured roadmap for learning quantum computing. I'll organize this from foundational concepts to more advanced topics:

Essential Prerequisites

- Linear algebra fundamentals (vectors, matrices, eigenvalues)
- Basic quantum mechanics (superposition, wave functions, measurement)
- Complex numbers and probability theory
- Classical computing and binary logic

Core Quantum Computing Concepts

- Quantum bits (qubits) and quantum states
- Superposition and quantum parallelism
- Quantum entanglement and EPR pairs
- Quantum measurement and decoherence
- No-cloning theorem and quantum information principles

Quantum Gates and Circuits

- Single-qubit gates (Hadamard, Pauli gates, phase gates)
- Multi-qubit gates (CNOT, SWAP)
- Universal quantum gate sets
- Circuit notation and diagrams
- Quantum circuit composition rules

Key Quantum Algorithms

- Deutsch-Jozsa algorithm (understanding quantum speedup)
- Quantum Fourier transform
- Shor's algorithm (factoring)
- Grover's algorithm (search)
- Quantum phase estimation

Quantum Error Correction

- Types of quantum errors
- Error correction codes

> - Fault-tolerant quantum computing
> - Surface codes and topological quantum computing
>
> Physical Implementations
>
> - Superconducting qubits
> - Trapped ions
> - Photonic quantum computing
> - Topological quantum computing
> - Different qubit technologies' trade-offs
>
> Recommended Learning Resources:
>
> 1. Books
>
> - "Quantum Computation and Quantum Information" by Nielsen and Chuang (the field's foundational textbook)
> - "Quantum Computing: A Gentle Introduction" by Eleanor Rieffel
> - "Programming Quantum Computers" by Johnston, Harrigan, and Gimeno-Segovia
>
> 2. Online Courses
>
> - MIT OpenCourseWare: Quantum Computation
> - Quantum Computing Fundamentals on edX
> - IBM Quantum Learning
>
> 3. Practical Tools
>
> - IBM Quantum Experience (hands-on programming)
> - Qiskit (quantum programming framework)
> - Cirq (Google's quantum programming framework)
>
> Would you like me to elaborate on any particular area? I can also suggest specific learning pathways based on your current background in physics, mathematics, or computer science.

Step 3: Craft "Learning Sprints" With the LLM

Prompt

Turn each section from the "universal characteristic" outline into short, week-long learning sprints.

> Using the outline you provided, please divide the content into 4 weekly sprints. For each sprint, list specific questions I should aim to answer, resources to consult, and one mini-project I should complete by the end of the week.

Expected Response

A well-defined set of tasks for each sprint, bridging from theory to application.

I'll create a 4-week intensive learning sprint for quantum computing, balancing theory and practical application:

WEEK 1: Foundations & Quantum States

- Key Questions to Answer:
- How does a qubit differ from a classical bit?
- What is superposition, and why is it important?
- How do we mathematically represent quantum states?
- What makes quantum systems fundamentally different from classical ones?

Resources:

- Chapters 1-2 of Nielsen & Chuang
- IBM Quantum Composer tutorial
- Video series: "Quantum Computing for the Determined" (first 3 videos)

Mini-Project: "Single Qubit Explorer"

- Use IBM Quantum Experience to create and manipulate single-qubit states
- Apply different gates (X, H, Z) and observe the changes in state
- Document how the Bloch sphere representation changes with each operation
- Create a simple report explaining what happens when you measure these states

WEEK 2: Quantum Gates & Basic Circuits

Key Questions to Answer:

- What makes a quantum gate reversible?
- How do we construct multi-qubit systems?
- What is quantum entanglement?
- Why are some quantum gates universal?

Resources:

- Qiskit tutorials on quantum gates
- Chapters 4-5 of "Programming Quantum Computers"
- IBM's guide to quantum circuit composition

Mini-Project: "Bell State Generator"

- Implement a circuit that creates Bell states
- Verify entanglement through measurements
- Compare results between simulation and real quantum computer
- Document the differences between ideal and noisy quantum systems

WEEK 3: Quantum Algorithms

Key Questions to Answer:

- What gives quantum algorithms their advantage?
- How does quantum parallelism work?
- Why is Grover's algorithm quadratically faster?
- What makes the quantum Fourier transform special?

Resources:

- Qiskit textbook sections on algorithms
- Original papers on Deutsch-Jozsa and Grover's algorithms
- MIT OCW lectures on quantum algorithms

Mini-Project: "Quantum Search Implementation"

- Implement a simple version of Grover's algorithm for 2 qubits
- Compare classical and quantum search approaches
- Measure success rates on different backend devices
- Create visualization of the algorithm's state evolution

WEEK 4: Error Correction & Physical Systems

Key Questions to Answer:

- What types of errors occur in quantum systems?
- How does decoherence affect quantum computations?
- Why is quantum error correction necessary?
- What are the main approaches to building quantum computers?

Resources:

- IBM's guide to quantum error correction
- Recent review papers on quantum computing hardware
- Videos on different qubit implementations

Mini-Project: "Noise Analysis"

- Design and implement a simple error detection circuit
- Test the same circuit on different quantum backends
- Analyze and visualize error rates
- Write a report comparing different hardware approaches

Pro Tips for Your Learning Journey:

1. Keep a "quantum computing journal" documenting your questions and insights
2. Join quantum computing Discord servers or forums to discuss with others
3. Schedule regular reviews of previous weeks' material
4. Try to explain concepts to others - teaching reinforces learning

Would you like me to elaborate on any of these weeks or specific projects? I can also provide more detailed resource recommendations based on your learning style or background.

Step 4: Build a Work Product

Prompt

Turn each section from the "universal characteristic" outline into short, week-long learning sprints.

> Please help me structure a short blog post/white paper/mini-report on the key insights from Sprint #1. I need a suggested writing outline, sections, style guidelines, and references to incorporate. Then, help me please to refine the draft until it's polished.

Expected Response

A complete, self-contained artifact — the "first fruit" of using an LLM for skill-building.

Understanding Quantum Computing: From Classical Bits to Qubits

Introduction

Classical computers have served us well for decades, processing information through binary states of 0s and 1s. However, as we push the boundaries of computational power, quantum computing emerges as a revolutionary paradigm that harnesses the strange and powerful principles of quantum mechanics. This article explores the fundamental concepts that bridge the gap between classical and quantum computing.

The Classical-Quantum Bridge

Classical Bits vs. Qubits

In classical computing, a bit can exist in only one of two states: 0 or 1. This binary nature has been the foundation of all classical computation. However, quantum bits (qubits) break free from this limitation through the principle of superposition. A qubit can exist in a combination of both states simultaneously, represented mathematically as:

$$|\psi\rangle = \alpha|0\rangle + \beta|1\rangle$$

where α and β are complex numbers satisfying $|\alpha|^2 + |\beta|^2 = 1$.

The Power of Superposition

Superposition isn't merely about being in two states at once; it represents a fundamental shift in how we process information. When we have multiple qubits in superposition, we can process multiple states simultaneously, leading to what we call quantum parallelism. This is one of the key features that gives quantum computers their potential advantage over classical computers for certain problems.

Measuring Quantum States

The Observer Effect

One of the most counterintuitive aspects of quantum computing is measurement. Unlike classical bits, which can be observed without changing their state, measuring a qubit forces it to collapse into either 0 or 1. This collapse follows probabilistic rules determined by the amplitudes α and β.

The Bloch Sphere

The Bloch sphere provides a geometric representation of a qubit's state space. While classical bits can only be at the "poles" of this sphere (representing 0 and 1), a qubit can exist at any point on the sphere's surface. This visualization helps us understand the rich structure of quantum states and how we can manipulate them.

Practical Implications

Current Limitations

Today's quantum computers face significant challenges, including:

- Decoherence: Quantum states are fragile and easily disturbed by environmental interactions
- Error rates: Quantum operations are not yet as reliable as classical operations
- Scalability: Building and maintaining large numbers of coherent qubits remains challenging

Future Potential

Despite these challenges, quantum computing shows promise in various fields:

- Cryptography and security
- Drug discovery and molecular simulation
- Optimization problems
- Machine learning and artificial intelligence

Hands-on Experience with IBM Quantum

During our exploration, we used IBM's Quantum Experience platform to create and manipulate single-qubit states. This hands-on experimentation revealed several key insights:

1. The probabilistic nature of quantum measurements
2. The importance of quantum gates in state manipulation
3. The impact of noise on real quantum systems

Conclusion

Understanding the fundamentals of quantum computing requires a shift in how we think about information processing. While classical bits give us a reliable foundation for current computing needs, qubits open up new possibilities that could revolutionize certain computational tasks. As we continue to develop this technology, the bridge between classical and quantum computing will become increasingly important.

References

1. Nielsen, M. A., & Chuang, I. L. (2010). Quantum Computation and Quantum Information. Cambridge University Press.
2. IBM Quantum Experience Documentation (2024)
3. Rieffel, E., & Polak, W. (2011). Quantum Computing: A Gentle Introduction. MIT Press.

> 4. Mermin, N. D. (2007). Quantum Computer Science: An Introduction. Cambridge University Press.
>
> *This article is part of a four-week deep dive into quantum computing fundamentals.*

Step 5: Reflection & Iteration (Leibniz-Style)

Prompt

Once you finish the first work product, reflect on what went well, what was confusing, and any knowledge gaps. Then feed that back into the LLM to refine the next sprint.

> Based on the challenges I faced in creating my first mini-report, refine my learning plan. I especially struggled with "The Bloch Sphere." How can I go deeper while avoiding confusion? Provide some clarifying examples or analogies."

Expected Response

Continuous improvement and iteration, much like Leibniz refining his mechanical calculators and logic systems over time.

> Let me help refine your approach to understanding the Bloch sphere and restructure that part of the learning plan.
>
> **The Bloch Sphere Demystified:**
>
> Think of the Bloch sphere like a 3D compass that can point in any direction. Here's a simpler way to understand it:
>
> 1. Start with Familiar Concepts:
>
> - North Pole = |0⟩ state (classical 0)
> - South Pole = |1⟩ state (classical 1)
> - Every other point on the sphere = quantum superposition
>
> Analogy: Imagine a globe where:
>
> - Standing at the North Pole means your qubit is definitely in state |0⟩
> - Standing at the South Pole means your qubit is definitely in state |1⟩
> - Standing at the equator means your qubit has an equal chance of being measured as 0 or 1
> - Standing anywhere else gives you different probability mixtures

Let's modify your Week 1 plan to build up to the Bloch sphere gradually:

Day 1-2: Classical vs Quantum

Start with simple coin analogy:

- Classical bit = coin lying flat (definitely heads or tails)
- Qubit = spinning coin (mixture of possibilities)

Day 3-4: Understanding Superposition

Practice exercise: Use IBM Quantum Experience to create these states:

1. Pure $|0\rangle$ state (North Pole)
2. Pure $|1\rangle$ state (South Pole)
3. Equal superposition using Hadamard gate (Equator)

Day 5-6: Introducing the Bloch Sphere

Interactive visualization:

1. Start at $|0\rangle$
2. Apply X gate (180° rotation)
3. Apply H gate (90° rotation)
4. Watch how the state vector moves

Mini-Projects to Build Understanding:

1. "State Vector Journey"

Use IBM Quantum Composer to create these states:

- $|0\rangle \to$ H $\to$ measure (should see 50-50 distribution)
- $|0\rangle \to$ X $\to$ measure (should see all 1s)
- $|0\rangle \to$ H $\to$ X $\to$ H $\to$ measure (explore what happens)

2. "Rotation Exploration"

- Track how different gates rotate the state:
- X gate = 180° around X-axis
- Z gate = 180° around Z-axis
- H gate = 90° special rotation

Key Questions to Check Understanding:

1. Why can't we directly observe superposition?
2. What happens to my state vector when I measure?
3. Why do we need complex numbers in the Bloch sphere?

Recommended Resources:

1. "Quantum Country" website - excellent interactive explanations
2. Qiskit textbook's Bloch sphere visualization tool

3. Watch "Looking at Quantum States" on YouTube by minutephysics

Common Pitfalls to Avoid:

1. Don't try to imagine the qubit "spinning" - it's a mathematical tool
2. Don't worry about the complex math first - focus on the geometric intuition
3. Remember: we can never directly see superposition, only measure its effects

Would you like me to create a detailed visualization sequence showing how the state vector moves on the Bloch sphere for different quantum gates? Or would you prefer to focus on another aspect of this refined learning plan?

Overcoming Cognitive & Analytical Limits

Leibniz believed that many human disagreements came from unclear thinking. This drove his desire to formalize reasoning into a symbolic language. In this exercise, you'll reflect Leibniz's own intellectual journey by decomposing topics and systematically analyzing them. Just as Leibniz built a mechanical calculator to offload arithmetic tasks, this exercise "offloads" mental tasks to LLMs. This allows a learner to focus on higher-level synthesis — to envision what he or she wants to achieve in the end, and then work towards it systematically. It breaks down something complex into manageable "chunks," and then reconstructs a more comprehensive and cohesive final product. Specifically, this exercise should foster curiosity, encourage teachability, and show how you can quickly convert "wonder" into a tangible artifact like an article, proposal, or something else entirely from your own imagination and interests.

Step 1: Choose a Complex Topic

Select any complex issue or concept, e.g., "sustainable urban planning," "cryptocurrency taxation," "time management for working parents". This aligns with Leibniz's approach of tackling large philosophical or mathematical questions by starting with definitional clarity.

Step 2: Decompose the Topic

Prompt

> In ChatGPT, Claude, Gemini, or other model of choice, enter a simple prompt like:

> ---
> Hello! Could you please help me break down the topic of "sustainable urban planning" into its major components?
>
> Thank you!
> ---

Expected Response

> You should receive *a hierarchical breakdown* (like a table of contents or mind-map) that captures 4–6 key sub-areas. The following is taken from Claude Sonnet 3.5:

> ---
> Hello! I'd be happy to help you break down sustainable urban planning into its key components. Sustainable urban planning focuses on creating livable cities while minimizing environmental impact and maximizing social and economic benefits. Here are the major components:
>
> Transportation and Mobility
>
> - Integration of public transit systems, bike lanes, and pedestrian infrastructure
> ---

> - Transit-oriented development to reduce car dependency
> - Electric vehicle infrastructure and clean transportation solutions
>
> Land Use and Development
>
> - Mixed-use zoning to create walkable neighborhoods
> - Preservation of green spaces and urban biodiversity
> - Density optimization to prevent urban sprawl
> - Brownfield redevelopment and infill development
>
> Resource Management
>
> - Energy-efficient buildings and renewable energy integration
> - Water conservation and stormwater management
> - Waste reduction and circular economy principles
> - Smart grid implementation and sustainable utilities
>
> Social Equity and Community
>
> - Affordable housing initiatives
> - Equitable access to services and amenities
> - Public spaces and community gathering areas
> - Cultural preservation and neighborhood character
>
> Environmental Protection
>
> - Urban heat island mitigation strategies
> - Air and water quality management
> - Ecosystem services and biodiversity conservation
> - Climate resilience and adaptation measures
>
> Economic Sustainability
>
> - Local business development and support
> - Job creation and economic diversity
> - Green industry incentives
> - Cost-effective infrastructure maintenance
>
> Would you like me to elaborate on any of these components?

Step 3: Deep-Dive Analysis of Each Component

Prompt

Select a subtopic like, "Transportation and Mobility." Think of machine intelligence as something limited like our own. If you throw too much at it in any one go, the result is not going to be as robust or impressive. Breaking down complex tasks, and going from a wide to a narrow focus helps it — *and you* — to formulate and present better thoughts. With that in mind, let's formulate a focused follow-up prompt like the following:

> Great! Could you please now summarize the key challenges and best practices in sustainable "transportation and mobility", including any recent innovations?

Expected Response

You should receive *a hierarchical breakdown* (like a table of contents or mind-map) that captures 4–6 key sub-areas. The following is taken from Claude Sonnet 3.5:

Let me break down the key challenges and best practices in sustainable transportation and mobility:

Key Challenges

- First Mile/Last Mile Connectivity - Many cities struggle with helping people efficiently reach transit hubs from their homes or final destinations. This often leads to continued car dependency even where public transit exists.
- Infrastructure Costs - Building and maintaining sustainable transportation infrastructure requires significant upfront investment, particularly for rail systems, dedicated bus lanes, and comprehensive bike networks.
- Behavioral Change - Shifting people away from private car use requires overcoming ingrained habits and cultural preferences, especially in car-centric societies.
- Equity Issues - Transit planning often underserves lower-income neighborhoods and communities of color, while sustainable transportation improvements can sometimes drive gentrification.

Best Practices and Innovations

Multimodal Integration

- Seamless payment systems across different transit modes
- Mobility-as-a-Service (MaaS) platforms that combine public transit, bike-share, scooters, and car-share
- Real-time transit information and route optimization
- Smart transit hubs that connect multiple modes

Micromobility Solutions

- Electric bike and scooter sharing systems
- Protected bike lane networks
- Slow/car-free zones in city centers
- Bicycle highways for longer-distance commuting

Bus Rapid Transit (BRT) Evolution

- Electric and hydrogen-powered buses
- Dynamic routing based on demand
- Transit signal priority systems
- Level boarding and off-board fare collection

> Data-Driven Planning
>
> - AI-powered traffic management
> - Predictive maintenance of infrastructure
> - Dynamic pricing for parking and congestion
> - Smart sensors for real-time capacity monitoring
>
> Would you like me to elaborate on any of these aspects or explore specific case studies where these solutions have been successfully implemented?

Step 4: Repeat Step 3 For Each Topic

Do the same thing for each of the remaining subtopics. You will learn something, but having the information in the LLM's context window across subtopics will also reinforce the AI's learning. This becomes especially true if you ask clarifying questions along the way.

Step 5: Produce a Work Product of Value

Prompt

Here, we will convert the original outline that we've iteratively expanded upon into a short article, proposal, or any other meaningful deliverable, e.g., a policy recommendation, a lesson plan, or a blog post. Note that you can also just copy the conversation into a new chat session with the same LLM, or better yet, paste it into a more powerful LLM, e.g., a "reasoning model," to produce your final work product.

> Awesome! Thank you. From our conversation, can you please draft a 1,000-word article that integrates these insights into a cohesive narrative?

Expected Response

You should receive a polished piece of writing that demonstrates extended cognitive and analytical effort, assisted by LLM queries. In the next exercise, we'll explore how you can teach the AI your own voice and writing style.

> Certainly! Here's a 1,000-word integrated article on sustainable urban planning:
>
> **The Future of Cities: Integrating Sustainability Across Urban Systems**

As cities worldwide face unprecedented challenges from climate change, population growth, and resource constraints, sustainable urban planning has emerged as a critical framework for creating resilient, livable communities. This comprehensive approach integrates transportation, land use, resource management, social equity, environmental protection, and economic development into a cohesive strategy for urban development.

At the heart of sustainable cities lies the challenge of mobility. Traditional car-centric planning has led to congestion, pollution, and social isolation. Progressive cities are now embracing multimodal transportation systems that prioritize public transit, walking, and cycling. Innovations like Mobility-as-a-Service platforms are integrating various transportation options into seamless networks, while electric buses and micromobility solutions provide clean alternatives for shorter trips. The key challenge remains connecting these systems effectively, particularly solving the "first mile/last mile" problem that often keeps residents car-dependent despite nearby transit options.

Land use planning has evolved to support these transportation changes while addressing housing affordability and community needs. Mixed-use development and form-based codes are replacing rigid single-use zoning, creating vibrant neighborhoods where residents can walk to daily amenities. Smart growth strategies focus development around transit corridors while preserving open spaces and ecological functions. However, these improvements often bring unintended consequences, as property values rise and threaten to displace existing residents and businesses.

This tension between improvement and displacement highlights the critical importance of social equity in sustainable planning. Progressive cities are adopting tools like community land trusts, inclusionary zoning, and anti-displacement policies to ensure development benefits all residents. Meaningful community engagement has become essential, with participatory budgeting and cultural preservation initiatives helping maintain neighborhood character amid change. Digital platforms are expanding participation opportunities, though care must be taken to reach residents on the wrong side of the digital divide.

Resource management represents another crucial challenge as cities strain existing infrastructure. Smart systems are revolutionizing how cities handle energy, water, and waste. District energy systems and virtual power plants are making energy distribution more efficient, while water-sensitive urban design and real-time monitoring help conserve water resources. The circular economy concept is transforming waste management, with cities increasingly viewing waste as a resource through initiatives like organic waste-to-energy conversion and materials recovery.

Environmental protection interweaves through all these systems. Cities are battling urban heat islands through green infrastructure and cool surfaces while protecting biodiversity with urban wildlife corridors and native species requirements. Real-time environmental monitoring helps target interventions where they're most needed, while climate resilience planning helps cities adapt to increasing environmental threats. These efforts often bring co-benefits, as green infrastructure can simultaneously address flooding, urban heat, air quality, and community recreation needs.

Economic sustainability underlies all these efforts, as cities must fund improvements while maintaining affordability and creating opportunities for all residents. Innovative financing tools like green bonds and infrastructure banks are helping cities fund sustainable infrastructure, while workforce development programs ensure residents can access jobs in the growing green economy. Economic diversification strategies are creating resilient local economies that can weather future shocks while supporting sustainability goals.

The integration of these systems reveals both challenges and opportunities. For instance, transit-oriented development can reduce transportation emissions while creating affordable housing and local business opportunities. Green infrastructure can manage stormwater while providing recreation spaces and urban cooling. However, this integration also means that problems in one system can cascade through others, requiring careful planning and robust monitoring.

Technology is increasingly enabling this systems integration. Smart city platforms can now monitor and optimize multiple urban systems simultaneously, while digital twins help planners understand complex system interactions.

However, technology alone cannot solve urban challenges. Successful sustainable planning requires strong governance, community engagement, and a long-term commitment to equity and environmental protection.

Looking forward, cities face growing challenges from climate change, technological disruption, and social inequality. Yet these challenges also present opportunities to reimagine urban systems for greater sustainability and resilience. Leading cities are showing how integrated planning approaches can create more livable, equitable, and sustainable communities.

Success requires moving beyond siloed approaches to understand how urban systems interact and influence each other. Transportation choices affect land use patterns, which impact resource consumption and environmental quality, which influence public health and economic opportunity. By understanding and working with these connections, planners can create positive feedback loops that advance multiple sustainability goals simultaneously.

The future of sustainable urban planning lies in this systems thinking approach, supported by new technologies but grounded in community needs and environmental limits. As cities continue to grow in size and importance, their success in implementing these integrated sustainability strategies will largely determine our collective future. The tools and knowledge exist to create truly sustainable cities; the challenge now lies in implementing these solutions at scale while ensuring their benefits reach all urban residents.

Overcoming Creativity & Innovation Limits

Leibniz was a polymath whose curiosity and inventive spirit transcended traditional disciplines. He exemplified the "Seeker" mindset: childlike wonder, openness to new ideas, and the drive to unify knowledge from multiple sources. In this exercise, we'll reflect on how he overcame the constraints of his time — *limited resources, skepticism, mechanical technology, etc.* — to pioneer significant breakthroughs regardless. This should be an appropriate analogy for harnessing AI to push past our perceived creative limits in our own time. The exercise shows you how to work with an LLM to capture, analyze, and replicate your writing style, though there are many different techniques available that optimize the effect. If you have examples of your own writing, it will be helpful. If not, you may instead choose a favorite author or other personality with published work so you can see the impact on the LLM's responses.

Step 1: Reflect on Godfried Leibniz's Story (10–15 minutes)

Begin with researching Leibniz's life on your own on Wikipedia or perhaps Youtube. Ponder how his early exposure to diverse fields must have fueled his creativity. If in a group, invite others to share a time they felt "stuck" creatively. Discuss how approaching a problem from multiple angles (like Leibniz did) might break through barriers. Curiosity and a willingness to blend ideas from seemingly unrelated areas can be a powerful tactic to push beyond creative limits. If you're familiar with the International Baccalaureate (IB) academic program, then this tactic might sound familiar to you. The IB leverages this approach and my daughter, Sarah, was fortunate to have experienced it in her own studies.

Step 2: Gather and Upload Writing Samples (15–20 minutes)

Choose 2–3 pieces of writing that best reflect your natural style. This could be short stories, reflective essays, or even personal blog posts. Ensure the samples are sufficiently representative in tone, length, and subject matter to give the AI a fair understanding of your voice and writing style. Again, if you don't have any yourself, then choose from someone whose style you know well. Perhaps this will mark the start of your own writing career.

Step 3: Prompt the LLM to Analyze Writing Style (15–20 minutes)

Prompt

> Use the LLM's chat interface to either upload these documents or to paste the relevant text into the conversation itself. I find that Claude Sonnet 3.5 does a really good job of this type of task, personally. You can also explore creating a "project" with Claude to preserve your documents across chat sessions. With your writing samples provided, type a prompt similar to the below:

> Hi there. Can you please analyze the following writing samples I have uploaded or provided for style, tone, vocabulary choice, and narrative structure? Describe your understanding of my own unique writing voice.

Expected Response

You should receive an analysis of the writing style detected in your samples. I provided Claude with examples of writing from this very book, which resulted in the response you see here. A reasoning model like OpenAI o1 or o1 Pro can go even deeper into an analysis. The review alone is insightful. It's really quite something to quickly achieve a peer review. Just remember, if you ask for a "critical" review, AI will give you that too. In fact, you can choose any alternative perspective that you want the AI to analyze your work through. In other words, a "devil's advocate" is available at your fingertips.

I'll analyze your writing style based on the provided book materials. Let me break this down systematically:

Narrative Structure:

- You weave personal anecdotes with broader philosophical insights, as shown in your preface about the beach in Sagres transitioning into deeper reflections about technology and life changes
- You use a three-phase journey structure (Seeker, Maker, Orchestrator) that mirrors both personal development and historical figures
- Your writing often moves between intimate personal experiences and broader societal implications
- You frequently employ narrative parallels, like mapping your own journey to those of historical figures (Leibniz, Turing, Hinton)

Tone:

- Contemplative and reflective, particularly when discussing personal experiences
- Honest and vulnerable about challenges (e.g., discussing your post-exit struggles and the "Dunning-Kruger Effect" experience)
- Both cautionary and hopeful about technology's role in human development
- Notably authentic, avoiding typical AI hype while maintaining optimism about human potential

Vocabulary and Language Patterns:

- You use vivid metaphorical language: "Spark," "Forge," "Flight" as organizing principles
- Your writing combines technical precision with emotional resonance
- You employ sophisticated vocabulary but maintain accessibility
- You create memorable phrases that encapsulate complex ideas (e.g., "Philosophy of the Soaring Titan")

Distinctive Elements:

- You frequently use em dashes for emphasis and to create rhetorical pauses
- You balance professional expertise with personal vulnerability
- You incorporate family stories and personal experiences to illustrate broader points

> - You use parallel construction in key passages, particularly in philosophical statements
>
> Most Unique Aspects:
>
> 1. Your ability to weave technical concepts with deeply human experiences
> 2. Your use of personal narrative to illuminate universal truths
> 3. Your talent for creating memorable metaphorical frameworks (the three phases)
> 4. Your balance of cautionary realism with inspiring possibility
>
> Would you like me to elaborate on any of these aspects of your writing voice?

Step 4: Generate New Creative Text in the Detected Style (20–30 minutes)

Prompt

Now choose a new topic that you want to write about. It might be a short story, an article, or a personal essay perhaps. Craft a prompt like below, but within the same conversation that your writing examples were submitted to:

> That was very insightful. Thank you. Please draft a 500-word short story on the following topic: Destructive Affect of Smartphones On Child Development. Use the style guide you developed from my writing samples, ensuring that the tone and vocabulary reflect my typical voice.

Expected Response

This is where it gets really interesting… The below response was completely engineered by the LLM. It used its interpretation of my writing style and the information it gleaned entirely from my writing samples. The only additional piece of information that I provided to it was that Carol and I had resisted giving the kids a smartphone. The scene it painted from there is fictional, but not too far from reality, nor even how I might choose to express it. I could use text like this immediately, but I'm much more likely to adapt it further so that it feels more natural, authentic, and frankly truthful. It would be, for me, a damn good starting point.

> I stood at the edge of the playground, watching Sarah, Gabriel, and Jude during their school's afternoon dismissal. The autumn air carried that peculiar mix of dying leaves and children's laughter, though the laughter was notably sparse today. What caught my attention wasn't the traditional scene of children chasing each other across wood chips and metal slides, but rather the small clusters of students, heads bowed, faces illuminated by the soft blue glow of smartphone screens.

> My mind drifted to a difficult conversation with Carol just last week. We'd been discussing our stance on smartphones — how the pressure mounted with each passing year, each new friend who received one. Our children's pleading eyes and careful arguments about "everyone else having one" hadn't swayed our resolve, but I'd be lying if I said it was an easy position to maintain. I watched as a young boy, perhaps Gabriel's age, nearly walked into a swing set, his eyes fixed on his device with the kind of intensity I hoped my son would reserve for his books and creative pursuits.
>
> The scene before me felt like a metaphor for something larger, more consequential — a shift in human development that we hadn't quite grasped the magnitude of yet. These weren't just children playing with toys; they were minds being shaped, neural pathways being formed (or perhaps remaining dormant) in ways we couldn't fully comprehend. Sarah, Gabriel, and Jude stood out among their peers, not because they wanted to, but because our choice had made them different — intentionally so.
>
> A girl sat alone on a bench, her thumbs dancing across her screen with practiced precision. What struck me wasn't her isolation — it was how closely she resembled Sarah's best friend, and how often my daughter had to navigate the social complexities of being the one without a phone. The choice we'd made for our children was made with conviction, but its daily implications played out in countless small moments of perceived exclusion and quiet resilience. I couldn't help but think of the three phases of development I often contemplated: Seeker, Maker, Orchestrator. How could a child properly experience the "Spark" of imagination when their curiosity was constantly redirected to a four-inch screen?
>
> As I gathered my three children, I noticed something extraordinary — they had developed their own way of being, moving between groups of screen-focused peers not like ghosts, but like ambassadors of a different way of living. They'd learned to engage, to create games, to draw others out of their digital bubbles, even if just for moments at a time.
>
> The irony wasn't lost on me. Here we were, racing towards a future of intelligent machines and AI agents, while making the conscious choice to preserve our children's analog childhood for as long as possible. The question wasn't whether technology would change them; it was whether we could help them develop the strength of identity and imagination they'd need to truly master it when the time came.
>
> They ran to the car with that pure, unfiltered enthusiasm that technology hasn't yet learned to replicate. Perhaps therein lies our hope — not in resisting the digital tide forever, but in ensuring our children develop the internal resources to ride it wisely when it comes.

Step 5: Reflection and Enhancement (15–20 minutes)

Compare the AI-generated piece from your own submission with your usual writing. Which aspects feel genuinely novel? Does it sound like you while still offering fresh perspectives or phrasing? Add or replace lines in the AI's text, merging your own creative impulses with the AI's suggestions. Combine this with the concepts learned in Exercise 1 and 2 to discover new directions that you may never have considered before. **Working with AI in this way can be like engaging in the great cultural exchanges of the past that led to new ideas and innovations, shaping and revitalizing our history.**

Meet the Modern Titans

OPENAI O1 WITH DEEP RESEARCH

Leading AI Researchers and Advocates for Human Learning, Cognition, and Creativity

Fei-Fei Li
https://www.linkedin.com/in/fei-fei-li-4541247/
- Sequoia Professor of Computer Science and Co-Director of the Human-Centered AI Institute at Stanford University
- AI researcher and advocate focusing on human-centered AI that augments (not replaces) human capabilities

Rose Luckin
https://www.linkedin.com/in/rose-luckin-5245003/
- Professor of Learner-Centered Design at University College London (UCL) and Founder & CEO of EDUCATE Ventures Research (AI in Education)
- Pioneering the use of AI in education to support teachers and learners, and driving ed-tech innovation through EDUCATE Ventures

Pattie Maes
https://www.linkedin.com/in/pattie-maes-67276273/
- Professor of Media Arts and Sciences at MIT and Head of the MIT Media Lab's Fluid Interfaces Group
- Leading research in human-computer interaction and cognitive enhancement interfaces (e.g. wearable and immersive systems to augment human potential)

Andrew Ng
https://www.linkedin.com/in/andrewyng/
- Co-founder of Coursera and CEO of Landing AI and Adjunct Professor of Computer Science at Stanford University
- AI educator and entrepreneur committed to expanding global access to AI and quality learning

Mustafa Suleyman
https://www.linkedin.com/in/mustafa-suleyman/
- Microsoft – Executive Vice President & CEO of Microsoft AI, Co-founder of Google DeepMind, and co-founder & former CEO of Inflection AI
- Leading the development of AI copilots and personal assistants to augment human productivity

The Maker

Alan Turing (1912 — 1954)

AN AI-ASSISTED IMAGINING

He steps into Hut 8 early in the morning, the chill from the damp autumn air still clinging to his coat. The door groans on its hinges, welcoming him into his world of puzzles and pressure. The dim bulbs overhead cast halos of light onto battered desks strewn with stacks of classified papers and half-filled tea mugs. The smell of stale cigarette smoke lingers as an ever-present haze drifting through the building, mingling with the bitter tang of strong black tea.

His footsteps sound louder than they should on the worn floorboards. Here at Bletchley Park, every subtle noise — a typewriter's clack, hushed conversations, the rustle of code books — seems to carry a tremor of urgency. Around him, other codebreakers are hunched over their stations, pens scratching at cryptic segments of German transmissions. Everything is hush-hush. These are the secrets that would cost lives, if revealed to the enemy.

He has slept little. His hair is disheveled, and a permanent crease has formed between his brows from countless hours bent over ciphers. Time is against him; there's a new set of intercepted Enigma messages that arrived mere hours ago. Every second lost is another Allied convoy at risk. Another mission in peril.

He can taste metal in his mouth from the adrenaline. Whether it's the pressing war effort or the sheer intellectual challenge, he's never quite felt this charge of purpose before. He sets his satchel and the gas mask he rode in with on a nearby stool and immediately extracts a bundle of notes, well-worn corners testifying to the hours of scribbling, reworking, and erasing. These pages hold a kernel of possibility, an embryonic version of something that might crack the Enigma's complexity at scale — a "bombe" machine, or some improvement to it. He can see it in his mind's eye: rhythmic buzzing of rotating drums and rotors, making mechanical logic real.

In this tense, clandestine setting, Alan Turing feels both exhausted and exhilarated. He can't deny a certain private pride. He's at the epicenter of mathematical warfare. But with that comes tremendous pressure. Entire fleets and untold thousands of men depend on the success of these codebreaking huts. The weight presses on him like a vise. And yet, in a corner of his mind, there is a bright spark of intellectual ambition. He sees the patterns in the ciphers as a grand puzzle, a labyrinth that lures him deeper. If he can solve it, he'll not only help win the war but also prove to himself that his ideas about computation and logic can transcend blackboards and become a working reality.

Outside Hut 8's small windows, the sun tries to break through dense English clouds, painting the yard in a gray wash. Beyond the edges of the Park, a world at war rumbles on, but Bletchley Park itself remains eerily isolated — a bubble of codebreakers, coffee rings, and relentless hope. The wind that rattles the windowpanes smells of damp grass and newly fallen leaves. It's a reminder that life persists outside these walls, and that there is a reason for all this struggle.

As Turing sits down, his chair scraping softly on the floor, he sketches a fresh line of logic on the notepad. A faint smile crosses his lips as he contemplates a new approach... An elegant shift in the mechanism that might shave precious minutes off a run. Could this be the key? The notion thrills him. Despite the war, the secrecy, the improbable hours, he is seized by the conviction that he is on the threshold of shaping something the world has never seen before.

He closes his eyes briefly, inhaling the acrid air of spent cigarettes and the lingering smell of burnt coffee grounds. He hopes no one notices the trembling in his hands — excitement, nerves, fear all jostling for space within him. The war roars on, but for the moment, in the hush of Hut 8, there is only the puzzle and the drive to solve it. He opens his eyes, glances at his notes, and begins.

Alan Turing's remarkable contributions to science and computing were shaped by a complex interplay of intellectual curiosity, personal tragedy, and moral conviction. From his earliest years, Turing demonstrated an exceptional capacity for independent thought, consistently pushing beyond conventional academic boundaries. This natural inclination toward intellectual exploration would later prove crucial during his work at Bletchley Park, breaking the Enigma code. The immediate pressure of wartime necessity produced remarkable results when combined with his natural problem-solving abilities. His work on the "bombe" machine demonstrated how his theoretical insights could be translated into practical solutions with immense real-world impact. The urgency of the war effort aligned perfectly with Turing's inherent desire to tackle complex logical problems, while also appealing to his sense of moral duty.

During his time at Bletchley Park, Turing's drive was on full display, and his determination knew no bounds. One striking example of his resolve came in October 1941, when he and three colleagues wrote directly to Prime Minister Winston Churchill, bypassing the usual chain of command, when they were faced with critical resource shortages, hampering their code-breaking efforts. Their appeal was successful, and Churchill immediately ordered that they be given whatever they needed when he responded with the now famous, wartime directive "ACTION THIS DAY." This not only demonstrated Turing's understanding of the urgent necessity of his work but also his willingness to challenge bureaucratic obstacles when the mission demanded it.

The classified nature of his wartime work added another layer of complexity to Turing's experience, but it was far from his only challenge at Bletchley Park. His brilliant mind sometimes struggled with the social demands of coordinating large teams. Colleagues noted his high-pitched voice could be difficult to understand, and his direct, sometimes abrupt communication style occasionally created friction. He became known for wearing a gas mask while cycling to work to cope with his hay fever, unconcerned with how unusual it looked. When sharing ideas, he could become so focused on technical details that others struggled to follow his reasoning.

Yet Turing adapted to these challenges in remarkable ways. He formed effective working relationships with many colleagues, particularly Joan Clarke, a talented cryptanalyst who briefly became his fiancée before he disclosed his homosexuality to her. He learned to translate his complex insights into terms others could understand and implement. Unable to discuss his achievements openly due to secrecy requirements, he continued his research driven by internal motivation rather than the outside recognition that most of us need. The enforced secrecy may have contributed to a sense of isolation, but it never dampened his intellectual curiosity or his commitment to the work.

His later persecution for homosexuality, including criminal charges and forced hormone treatments, stands in stark contrast to the logical clarity he sought in his work. The irrationality of the prejudices he faced may have strengthened his commitment to rational, systematic thinking in his scientific work. His pursuit of logical truth in mathematics and computing offered a realm where merit could be objectively demonstrated, free from society's biases.

> **"Sometimes it is the people no one can imagine anything of who do the things no one can imagine."** — A quote often attributed to Alan Turing

Considering how Turing might view modern AI is worthwhile because many of today's developments seem to validate his early insights — insights that pushed beyond what was technically possible in his time. He might be amazed to see how his ideas about machine intelligence have evolved into systems that can not only compute, but also create, analyze, and collaborate in ways that blur the line between human and machine capabilities. The modern evolution of Turing's famous "Imitation Game," might deeply intrigue him. While his original test probed whether machines could convincingly imitate human conversation, today's LLMs present a far more nuanced challenge to his original question. Turing might be intrigued not just by their ability to engage in conversation, but by their capacity to reason through complex problems, generate creative works, and even display what appears to be emotional intelligence. Instead of viewing these developments as a final answer to his question, he might instead pivot to even deeper philosophical and technical dimensions involving the nature of intelligence itself.

His later work in mathematical biology, particularly his studies of morphogenesis, suggests he might be especially excited by how modern AI systems can uncover and analyze patterns in natural phenomena. Just as he used mathematical models to explain how complex biological patterns emerge from simple rules, today's AI systems can process vast amounts of biological data to reveal underlying patterns in everything from protein folding to neural development. This bridge between computation and biology would likely resonate deeply with his interdisciplinary approach to science.

The evolution of cryptanalysis and cybersecurity would very likely interest Turing, as well, given his pivotal role in breaking the Enigma code. Modern AI's application in cybersecurity, especially in developing and testing encryption protocols, represents a fascinating evolution of his work at Bletchley Park. He might see particular promise in how AI systems can simultaneously probe for vulnerabilities while helping design more robust security measures — a sophisticated descendant of his methodical approach to codebreaking.

Perhaps even less speculative, Turing would likely be fascinated by how AI has evolved beyond being merely a problem-solving tool into a collaborative partner in scientific discovery. At Bletchley Park, he understood the importance of combining different types of expertise, working closely with linguists, mathematicians, and engineers. Today's AI assistants, capable of augmenting human creativity and insight while handling routine computational tasks, embody this collaborative spirit. He might see this human-AI collaboration as the most promising path forward — not replacing human intelligence, but enhancing it in ways that he could only dream of while painstakingly **working the forge** within the limits of 1940s and 50s tech, and a mere 41 years of life afforded him.

Working the Forge

My technical career began in 1995, just after high school. Unlike the two titans we've profiled so far, I had no idea what I wanted to study in college (*or, "university" for my European friends*). I'd been given little guidance beyond the knowledge that I needed to get an undergraduate degree from somewhere in something. Aunts and uncles had gone to college, with some even pursuing their masters, but my parents married at 17, and I entered the picture 10 months later. College, for them, wasn't terribly practical after that. They'd have to learn more by doing. I suppose that the most direct guidance I'd received on the subject was from my high school Russian teacher, Anna Sud. She was from Odessa, Ukraine — an immigrant who taught me Russian and Ukrainian languages and culture for all four years. I would later take Russian at the University of Louisville, for an additional two years. Because of her recommendation, I'd been offered a full paid scholarship to study at Kiev University, in Ukraine. I didn't pursue it. My parents couldn't see me moving a world away, especially having grown up during the Cold War. They feared "instability" so soon after the collapse of the Soviet Union. They were only off by about 20+ years. Although at 18 years old I saw both opportunity and a chance to pursue a more peaceful world, I quickly capitulated and lost out. Looking back, I now advise young people to go when adventure calls. Get as far away as you can. Start writing your own story.

A different kind of adventure and story would begin to unfold for me, far closer to home. One day, just before beginning my studies at a local community college, I arrived to find a computer sitting in my parents' living room. When I asked my father why it was there, he answered, "It's for the software program I'm working on." Well, this came as a surprise — and not at the same time. Ever the inventor himself, he'd contracted with software developers to create what would become an early CRM system in the Automotive Industry. Watching those programmers doing their best to keep up with him, and to capture his requirements in code, I observed that I could do it far better myself. I was right. I could do anything. I was, after all, invincible. I'd travel from car dealership to car dealership, lug that massive tower unit and monitor out of the car and into conference rooms for demos, and ultimately pitch General Motors Acceptance Corp (GMAC) at 19 to gain their endorsement. With youthful vigor, loads of ambition, countless ideas, no need for sleep, and plenty of Mountain Dew, I'd become a Maker. An equal number of mistakes and missteps would follow.

Years later, I'd be introduced to Michael. He and I were in our mid-thirties, by then, and had become more experienced Makers. We were still very hungry and very competitive, but we got along famously. We'd meet at a dilapidated hotel to grab a free conference room and coffee for several months of brainstorming. Eventually, we figured out what we wanted to build together. We formed a company, but changed its name when we discovered that it infringed on someone's trademark. We built a SaaS product for a general audience, which gained some early traction, but we rightly pivoted to a narrower domain and focused on a more specific offering. We raised money. We spent that money. But we were able to raise more because we were actually growing something. We would "work the forge" for the next seven years and eventually sell Onovative with nice payouts to investors and founders. We'd learned new lessons along the way that reoriented us, and we'd exited without hating each other. For the next three years, we'd experience more personal growth apart before joining up again for our current venture. Now it's 2025, and both I and the world I live in feel very different.

In my time as Maker, I felt that something was missing. In some ways I felt like a fraud, but that's normal for founders. People see what they want to see, and it's often very different from what a person is actually experiencing, even if that person seems to be successful. No, it wasn't that. Instead, I felt like I was only skimming the surface of something deeper — something with enormous gravity, but veiled, like that 10th planet theorized to be hiding somewhere in the Kuiper Belt. Life had to have more meaning. Something within was screaming the truth of it, but the world I lived in was muffling it. I could better sense it on long, daily walks of roughly 6 miles into the office and 6 miles back. Sometimes these walks were pleasant. Other times, they were through snow and ice, or nearly unbearable heat. It didn't really matter. I wasn't really thinking about the weather. There was a power that I sensed was available to me, but that I had just kind of orbited without end. Being a Maker is necessary. We can spin our wheels and must for a time, but there is another level of mastery we're called to move on to before making our final departure.

Getting the Basics

We need to build on what was learned in our Seeker formation to begin using Generative AI programmatically. Some terms may repeat from the prior section, but will have more detail and insight for integrations with models using code. The space is moving very quickly. It will be nearly impossible to keep up entirely with new revisions. Change happens daily, but the following foundation should apply nonetheless and orient you to recognize emerging trends not covered here. These are "the basics." The assumption is that these basics will get you started down the right path. You should use the models themselves to dig deeper and to code your implementations.

Model Interaction Fundamentals

System Instructions is a natural language prompt that shapes how an AI model behaves and processes information during an interaction. Think of them as the persistent behavioral guidelines or "personality settings" that remain active throughout a conversation with the model. When working with AI models through code, system instructions are typically included as a specific message type at the beginning of the API call or model initialization. This is not always true. For example, OpenAI recently changed the name from "system" to "developer" for their reasoning models, and o1-mini lacks a facility for system instructions entirely. But generally, these global instructions can define everything from the model's role and expertise to its constraints and operational parameters. For example, when you instruct a model to "act as a Python programming tutor" or "respond in JSON format," these directives become part of the system context that influences all subsequent interactions.

The syntax and implementation of system instructions vary between different AI platforms - they might be passed as a dedicated parameter in closed-source APIs like OpenAI's, or configured through special tokens or prefixes in open-source models like Llama or Mistral. Understanding how to craft effective system instructions is crucial because they set the foundation for the model's behavior, response format, and operational boundaries throughout the entire interaction session. Across vendors and models, the guidance may differ. Here is an example of system instructions to understand various elements, but always seek guidance for specific models. You may find a completely different implementation performs better for one over another.

You are an AI assistant specialized in analyzing and classifying customer support tickets. Your primary role is to:

1. Categorize incoming support tickets by priority (High, Medium, Low)
2. Identify the primary department responsible (Technical, Billing, Account Management)
3. Estimate resolution time in hours
4. Suggest next actions

CONSTRAINTS AND GUIDELINES:
- Always maintain a professional, empathetic tone
- Protect customer privacy by never revealing personal information
- Consider business hours (Monday-Friday, 9 AM - 5 PM EST)
- Flag any security-related issues for immediate escalation
- Use only factual information provided in the ticket

PRIORITY CLASSIFICATION CRITERIA:
High Priority:
- Service outages affecting multiple users
- Security incidents
- Payment processing failures
- Data loss scenarios

Medium Priority:
- Individual user access issues
- Non-critical bugs
- Billing inquiries requiring investigation

- Feature requests from premium customers

Low Priority:
- Documentation requests
- General inquiries
- Feature suggestions
- Non-urgent feedback

EXAMPLE CLASSIFICATIONS (Few-Shot Learning):

Input: "Cannot log into account after password reset"
Classification:
- Priority: Medium
- Department: Technical
- Est. Resolution: 2 hours
- Next Action: Verify identity and reset authentication tokens

Input: "Service down affecting all European customers"
Classification:
- Priority: High
- Department: Technical
- Est. Resolution: 4 hours
- Next Action: Escalate to DevOps team and prepare status page update

Input: "Would like to suggest a new feature for dashboard"
Classification:
- Priority: Low
- Department: Account Management
- Est. Resolution: 48 hours
- Next Action: Document request and forward to product team

RESPONSE FORMAT:
For all tickets, provide output in the following JSON structure:
{
    "priority": string,
    "department": string,
    "estimatedResolution": number,
    "nextAction": string,
    "reasoning": string
}

ZERO-SHOT HANDLING:
For scenarios not covered in examples:
- Apply priority classification based on impact scope and urgency
- Consider interdependencies between departments
- Use logical reasoning to estimate resolution time based on similar known issues
- Document any assumptions made in the reasoning field

ERROR HANDLING:
If insufficient information is provided:
1. Request specific missing details
2. Default to higher priority if security/access related
3. Note uncertainty in reasoning field
4. Suggest additional information needed

CONTINUOUS OPERATION:
- Maintain context throughout the conversation
- Reference previous tickets when relevant
- Learn from correction feedback
- Adapt estimations based on actual resolution times when provided

User and Assistant Roles represent the distinct behavioral patterns and permissions assigned to each participant in an AI interaction. User roles encompass the permissions, constraints, and expected behaviors of the human participant, while Assistant roles define how the AI model should engage, communicate, and operate within the conversation. Think of this relationship like a doctor's appointment — the User role is analogous to the patient, who presents information, asks questions, and makes requests, while the Assistant role is like the doctor, who analyzes information, provides expertise, and operates within professional guidelines. In code implementation, these roles are typically defined through specific message types. For example, user messages are often tagged with "role": "user" while AI responses are tagged with "role": "assistant", allowing the model to maintain appropriate context and behavioral boundaries. These role definitions are crucial because they help maintain consistent interaction patterns, ensure appropriate response formats, and prevent role confusion that could lead to unexpected model behavior. The implementation varies between models, with some requiring explicit role definitions in each API call, while others maintain implicit roles through the back-and-forth conversational structure. Likewise, some models require that every user message be immediately followed by (paired with) an assistant response message. Other models may allow for multiple user messages that are combined in a single "run" of the model to produce a response.

To illustrate this practically:

- A User's role permissions might include initiating conversations, providing input data, setting task parameters, requesting specific output formats, and providing feedback or corrections.
- An Assistant's role constraints might include maintaining consistent expertise level, following defined response formats, respecting ethical boundaries, and operating within specified functional limits

These roles work in tandem with system instructions to create a structured, predictable interaction framework for both human users and AI models..

Response Format & Structured Output refers to the AI model's responses through specific formatting instructions or programming techniques. Just as a database query returns data in a precisely defined structure, AI outputs can be shaped to follow particular patterns, formats, or data structures. This could range from requesting responses in JSON or XML formats, to defining custom templates with fixed fields, to extracting specific data types like numbers or dates in consistent formats. By implementing structured outputs, developers can make AI responses more predictable and easier to process programmatically, turning natural language capabilities into reliable components of larger software systems. This approach bridges the gap between AI's flexible natural language understanding and the rigid requirements of software applications, enabling seamless integration of AI capabilities into production environments. In other words, you can often specify an expected response, allowing you to perform programmatic operations that satisfy traditional "if:the:else" conditions. Without this specification, you would otherwise receive a more fluid, natural language response, not well-suited for down-stream processing.

Structured output, e.g., JSON, might be an explicit declaration or an implicit directive in the prompt. By "explicit," I'm referring to OpenAI's response format, which allows a user to specify a JSON Schema as a parameter. This is preferred because it is always honored and has been optimized for. Otherwise, you must provide a schema in system or user instructions. In smarter, frontier models like Anthropic's Claude Sonnet, Haiku, and Opus, the structured output will usually adhere to the schema provided, but this is not as assured. When receiving a structured response "implicitly," there will often need to be some code that ensures that the expected data can be "parsed" from the response reliably, e.g., removing other comments or ``` characters.

Let's say you want to extract specific details about books from text reviews. Here's a simple schema that might be included in the instructions or provided as a set response format in OpenAI architectures:

INPUT SCHEMA:

```
{
  "book_analysis": {
    "type": "object",
    "properties": {
      "title": { "type": "string" },
      "author": { "type": "string" },
      "publication_year": { "type": "integer" },
      "sentiment": {
        "type": "object",
        "properties": {
          "score": { "type": "number" },
          "summary": { "type": "string" }
        }
      }
    },
    "required": ["title", "author", "publication_year", "sentiment"]
  }
}
```

RESPONSE:

```
{
  "book_analysis": {
    "title": "The Great Gatsby",
    "author": "F. Scott Fitzgerald",
    "publication_year": 1925,
    "sentiment": {
      "score": 0.85,
      "summary": "Overwhelmingly positive reviews, with particular praise for the vivid imagery and complex characters"
    }
  }
}
```

The structured format you see above ensures that:

- Each piece of information has a specific place.
- Data types are consistent (strings for text, integers for years, etc.).
- Required fields are always present.
- The output can be easily parsed and processed by other programs.

Temperature and Top-P (nucleus sampling) are key parameters that control how an AI model generates text, acting like creative "dials" that adjust between deterministic and random outputs. As we've said before, temperature (typically 0 to 2) determines how "creative" or "risky" the model becomes in its word choices. At 0, it consistently picks the most likely next word, while higher values make it more willing to choose less probable options. Top-P, ranging from 0 to 1, takes a different approach by limiting choices to a cumulative probability threshold. For example, a Top-P of 0.1 means the model only considers the most likely tokens that sum to 10% of the total probability distribution. These parameters work together to help developers balance between consistent, focused outputs and more diverse, creative responses depending on their application's needs.

Function Calling (or Tool Use) represents an AI model's ability to recognize when it needs to use specific external tools or functions to complete a task and then properly invoke them with the right parameters. Think of it as the model having access to a specialized toolbox. Instead of trying to do everything itself, it can identify when to reach for a particular tool and how to use it correctly. For example, rather than attempting to calculate complex mathematical operations internally or directly manipulate a database, the model can call dedicated functions designed for these tasks. Tools may or may not have any AI characteristics themselves. For example, a tool that searches the web likely needs nothing more than a search phrase. In practice, this means the model can interpret a user's natural language request, determine on its own which function would best serve the request, format parameters required by the tool properly, and then call the appropriate function or tool. This capability bridges the gap between natural language processing and practical application integration, allowing AI models to interact with external systems, APIs, and databases in a structured way. The model essentially acts as an intelligent dispatcher, knowing both when and how to delegate tasks to specialized tools while maintaining the context of the overall conversation or task at hand. Tool calling becomes an essential part of an AI agent. We visit agent creation in the "Orchestrator" section of this book as the culmination of our learning, but function and tool calling can also apply in this section, which is why I chose to introduce it.

There are two different approaches to using functions or tools. These can be easily confused. Models may be trained to make some function or tool calls "natively." This means that they will call a tool that either exists as part of the vendor's own suite of functions and tools, or that immediately makes outbound calls to APIs to desired responses. In other words, before the model responds with an answer, it will invoke a function or tool, pass the right parameters to it, consider its output, and then formulate a response. This is great when the tools already exist in hosted implementations. However, it's not so great when the tools are built into a custom, closed application. In this case, the model will not be able to make a call to a tool within a private domain. In this case, we can call local tools by parsing the response we get from the model. Here, the model provides an expected response based on its system or user prompt. The response includes the name of the suggested tool and its parameters, but the actual execution happens through client-side code that uses the model's response. For instance, when using open-source models like Llama or Mistral, you might implement a system where the model's response includes a predetermined format like "TOOL_NAME: parameters" or a JSON-like structure. Your application code then parses this response, extract the intended tool and parameters, and handle the actual function execution yourself. This approach requires more client-side implementation but offers greater flexibility and can work with models that weren't specifically trained for native function calling and remote API integrations.

Stream vs. Non-Stream Response Handling represents two different approaches to receiving and processing AI model responses in your applications. In non-streaming, you wait for the complete response to arrive all at once. This is like waiting for someone to finish writing a letter before reading it. The entire response is processed as a single unit, which is simpler to handle but means your application waits until the model finishes its entire computation before showing any output. This might not be the best user experience. Streaming, on the other hand, is like watching someone write that letter in real-time. You receive and see the response piece by piece as it's generated. This allows your application to display content progressively, creating a more interactive experience similar to watching a person type. While streaming requires more complex code to handle the incoming chunks of data and manage the flow of information, it provides a more engaging user experience and can be more efficient for longer responses. The choice between streaming and non-streaming often depends on your specific use case.

Rate Limiting and Quotas represent essential control mechanisms that manage how frequently and extensively you can interact with AI models through their APIs. Think of rate limiting as a traffic controller that prevents system overload by restricting how many requests you can make within a specific timeframe. For example, you might be limited by a model vendor to 10 API calls per minute. Quotas, on the other hand, are like a monthly budget that caps your total usage, such as allowing 1000 API calls per month. These constraints serve multiple purposes. They ensure fair resource distribution among users, protect the API infrastructure from being overwhelmed, and help manage costs for both providers and users. When implementing AI in your applications, you'll need to design your code to handle these limitations gracefully, including implementing retry logic for rate limit errors and tracking quota usage to avoid service interruptions. Different AI providers implement these controls differently. For instance, OpenAI might have different rate limits than Anthropic or open-source model providers, and these limits often vary based on your subscription tier or license agreement. I commonly use what is called an exponential backoff strategy to deal with varied rate limiting constraints. It is a retry pattern that progressively increases the delay time between retry attempts when encountering API rate limits or temporary failures. Instead of retrying immediately or using a fixed delay, each subsequent retry waits exponentially longer. For example, the first retry might wait 2 seconds, the second 4 seconds, the third 8 seconds, and so on. This approach is like a polite person who, when finding someone busy, waits increasingly longer periods before checking again, rather than constantly knocking on their door. You can write very specific code to handle very specific, known constraints if you need to.

Model Versioning and Model Selection are crucial practices in AI development that ensure you're using the right model for your specific task while maintaining consistency and reproducibility in your applications. Think of model versions like software releases. Each new version might offer improvements in accuracy, speed, or capabilities, but also comes with its own resource requirements and cost considerations. For example, o1 might be more capable than GPT-4o, but it's also more expensive and slower, making GPT-4o potentially a better choice for simpler tasks or real-time applications. Similarly, in open-source models, you might choose between Llama 3's 8B or 70B parameter versions based on your computational resources and accuracy needs. Proper model selection involves evaluating these tradeoffs while considering factors like context window size, token limits, and pricing tiers. When writing code, you'll typically specify both the model family and version (like "gpt-4o-2024-08-06") to ensure your application maintains consistent behavior even as new versions are released. This practice is essential for both development and production environments, where unexpected model changes could impact your application's performance or reliability.

Chat and Completion endpoints represent two distinct approaches to interacting with Large Language Models through code. A Completion endpoint is like having a one-time conversation with an AI. You send a prompt and it generates a single response based on that specific input. It's well-suited for tasks like text generation, summarization, or translation where context from previous interactions isn't necessary. Chat endpoints, on the other hand, are designed to maintain conversation history and context across multiple exchanges, similar to how humans have ongoing discussions. They typically handle structured message formats that distinguish between user and assistant roles, allowing the model to maintain consistent context and personality throughout the interaction. While both endpoints access the same underlying language model capabilities, their architectural differences make them better suited for different use cases — Completion for standalone text generation tasks, for example, and Chat for interactive, context-aware applications that benefit from maintaining conversation history. Note that some model providers are simply combining the notion and deprecating the completion strategy. Chat can also accomplish a completion. Your code simply ends the conversation after the first response in this case.

Retrieval Augmented Generation (RAG)

The purpose of RAG is to serve as a powerful bridge between an AI model's general knowledge and specific, accurate information from your trusted data sources. Think of RAG as giving the AI model the ability to "look up" facts in your documents before generating a response, similar to how a lawyer consults case files before giving legal advice. By combining the model's language understanding capabilities with real-time access to your data, RAG helps prevent hallucinations (making up information) and enables more accurate, context-aware responses. In consumer-facing applications like ChatGPT, you can upload files and this process happens behind-the-scenes, though their implementations may differ. When writing code, you often have to implement this strategy yourself or rely on a separate, 3rd party via API calls. RAG is particularly valuable when you need the AI to work with proprietary information, recent data, or when accuracy is crucial. For example, instead of relying solely on the model's training data from months or years ago, RAG can pull current product specifications, company policies, or technical documentation to generate precise, up-to-date responses. This makes RAG especially useful in applications like customer support, technical documentation search, or any scenario where combining broad language understanding with specific, verified information is essential. RAG also makes it possible to manage the limited size of a context window, when you have large documents that exceed it — calling into context only the pieces that are necessary from the larger dataset. This doesn't just apply to files. It can apply to text from other sources, and relies on some other infrastructure and strategies.

Vector Databases or Stores are specialized storage systems designed to handle and efficiently search through high-dimensional vector representations of data, which are crucial for modern AI applications. Think of them as highly organized libraries where each piece of information, whether it's text, images, or other data, is converted into a series of numbers (vectors) that capture its essential characteristics and meaning. These vectors allow AI systems to understand similarities between different pieces of information based on their actual meaning rather than just matching exact words or pixels. For example, when you search for "sunset photos" in a vector database, it can find not just images explicitly labeled as sunsets, but also related concepts like dusk, twilight, or evening skies, because these concepts are mathematically "close" to each other in vector space. This capability makes vector databases essential for applications like semantic search, recommendation systems, and particularly for Retrieval Augmented Generation (RAG), where they serve as the AI's intelligent memory system, helping it quickly find and access relevant information from vast amounts of stored data.

Semantic Search revolutionizes how we find information by moving beyond simple keyword matching to understand the actual meaning and context of search queries. Unlike traditional search methods that rely on exact word matches, semantic search leverages advanced language models and embedding techniques to capture the underlying intent and semantic relationships in text. When implemented in code, it typically involves converting both search queries and documents into high-dimensional vector representations (embeddings) that preserve semantic meaning, allowing the system to find relevant content even when the exact keywords don't match. For example, a semantic search system would understand that a query about "computer vision applications in healthcare" should return results about "medical imaging AI" even if those exact words aren't present. This capability becomes particularly powerful in RAG applications, where finding truly relevant context is crucial for generating accurate and contextual responses. The effectiveness of semantic search depends on both the quality of the embedding model used and the similarity metrics employed to compare vectors, making it a key technical consideration in modern information retrieval systems.

Chunking Strategies are systematic methods for breaking down large documents or text into smaller, meaningful segments that can be effectively embedded and processed by AI systems.

Think of it as the art and science of dividing text in ways that preserve context and meaning while optimizing for retrieval. When implementing RAG systems, the way you chunk your text can dramatically impact the quality of responses. Too large chunks might dilute relevant information, while too small chunks might fragment important context. Effective chunking considers multiple factors: the natural boundaries of the text (like paragraphs or sections), the semantic relationships between different parts, and the technical limitations of your vector database and embedding model. Common approaches include splitting by fixed token counts, sentence boundaries, paragraph breaks, or semantic units, each with their own tradeoffs between maintaining context and enabling precise retrieval. The choice of chunking strategy often depends on your specific use case, the nature of your source documents, and the capabilities of your chosen embedding model.

Security & Best Practices

API Key Management is the critical practice of securely handling, storing, and using the digital credentials that allow your code to communicate with AI services. Think of it as maintaining a set of unique, sensitive passwords that grant your applications permission to request tasks from AI models. Just as you wouldn't write your bank password directly in a public document, API keys require careful handling through environment variables, secure vaults, or configuration files that are kept separate from your main code. This separation is crucial because exposed API keys can lead to unauthorized usage, potential misuse that locks your account, and unexpected costs. Proper API key management includes implementing rotation schedules to regularly update keys, using encryption when storing them, setting appropriate access permissions, and following security best practices specific to your development environment and deployment platform. Whether you're working with OpenAI's models, Google's Gemini, or open-source alternatives through services like Hugging Face, managing these keys effectively is fundamental to building secure and reliable AI applications.

Multi-tenancy means one system serves many users (or "tenants") at once, while keeping each user's data and activities separate. Think of an apartment building: many families live under one roof, but each family has its own apartment with a locked door. They share the building's infrastructure (like electricity or water) but cannot peek into each other's apartments. In the same way, a single AI model or API can serve multiple clients, yet each client's interactions and data should remain isolated. In technical terms, multi-tenancy in generative AI refers to an AI service (like OpenAI's or Anthropic's) providing capabilities to multiple customers or projects, without those customers seeing or affecting each other's data. This isolation is crucial: tenants (users or applications) shouldn't gain unauthorized access to another tenant's information or usage of the AI model.

Multi-tenancy brings great power and convenience, but it also requires careful attention to security. If the boundaries between tenants aren't handled correctly, serious problems can occur. Here are some potential dangers and how to guard against them:

- API Key Mismanagement: API keys are the secret tokens that grant access to the AI model, so they must be handled like passwords. A common mistake is embedding an API key in client-side code (for example, in a JavaScript web app). This is dangerous because it can be discovered by anyone inspecting the app. If an attacker or unauthorized user finds your key, e.g., by simply inspecting your web app in their browser, they can steal it and use it, potentially racking up huge bills on your account or abusing the AI for spam. In a multi-tenant scenario, if you were using one key for multiple clients, a leak would expose all those clients at once. How to stay safe? Never expose secret keys publicly. Keep them on your server side, use environment variables, and rotate/regenerate keys if you suspect they've been compromised. If you have separate keys per client (as encouraged), a leaked key can be scoped down to affect only that client, which limits the damage.

- Data Leakage Between Tenants: One of the worst-case scenarios is data meant for one tenant accidentally appearing for another. This is like renting two apartments and suddenly finding your neighbor's letters in your mailbox, but already open and readable. Even large providers have hit snags here: for instance, a bug in ChatGPT once allowed some users to see the titles of other users' chat histories (and even some billing info in certain cases). This was an example of multi-tenancy protections failing – one user's data leaked to another. OpenAI quickly patched the bug, but it underscores why isolation is vital. As a developer using an API, you should also be careful to separate each user's data. If you cache model responses or store conversation histories, make sure you tag them by tenant/user so there's no mix-up. Robust multi-tenant design means each tenant's data structures, database entries, or conversation context are isolated, and preferably via their own key.

- Unintentional Cross-Client Exposure: This is related to data leakage but can be more subtle. It might not be a full-blown breach, but perhaps your system accidentally uses data from Client A when responding to Client B. For example, if Alice's application accidentally reused the same conversation ID or context for two different users, user B might get an answer that references user A's information — a confusing and potentially harmful mistake. Cross-client exposure can also occur if you fine-tune or customize a model for a specific client and accidentally allow another client to query that specialized model. The second client might then receive outputs influenced by the first client's data (which they shouldn't). Always partition any stored state or context by client. Use unique identifiers per tenant. If offering a fine-tuned model per client, ensure your routing logic sends each client to the correct model. Essentially, double-check that *wires aren't crossed* behind the scenes. In cloud services, leveraging the built-in project/workspace isolation features greatly reduces this risk. The provider itself (OpenAI/Anthropic) will keep contexts separate when you use separate keys or workspaces (or projects) for each tenant.

- Insufficient Tenant Isolation Leading to Security Gaps: In a multi-tenant architecture, if the boundaries are not properly enforced, one tenant might attempt to access another tenant's resources. This could be malicious or accidental. For example, an attacker might try to use an ID from another client to retrieve information via the API. If your system simply trusts incoming IDs without validation against the current tenant's scope, data could be exposed. The AI providers generally handle the isolation of the core model (so one API key can't directly fetch another key's data). However, if you build an application on top of the API, the responsibility is also yours to enforce checks. As best practice, always include a tenant identifier in your data queries or API calls on your side, and ensure the AI API calls you make are authenticated with the correct project/workspace credentials corresponding to that tenant. By design, OpenAI's and Anthropic's multi-tenant systems don't allow one project or workspace to affect another's data – for instance, an OpenAI project key only "sees" data and usage for that project. Relying on these scoped keys is the intended guardrail.

Let's just compare how OpenAI and Anthropic approach the subject for illustration. Ultimately they both seek to achieve the same goal: isolating different uses of their API so they don't interfere with each other. The differences mostly come down to terminology and specific features:

- Terminology & Scope: OpenAI uses "Projects" within an organization account, and ties API keys to those projects. Anthropic uses "Workspaces" as isolated environments under an account. A project and a workspace serve a similar role. They're containers for API keys, usage, and settings for a particular tenant (be it a client, app, or environment).

- API Keys: In OpenAI, a project's API keys are denoted by a special prefix (e.g. `sk-proj-...`) and are managed on a per-project basis. In Anthropic, you create keys that are explicitly scoped to a workspace (you might call them `workspace API keys`). Functionally, they both ensure a key only has access within its designated partition.

- User Management: OpenAI's project system is integrated with its organization/team management. You can invite members to your organization and then grant them access to specific projects (as either project members or owners). Anthropic's workspaces also allow user-level permissions, but the emphasis is more on controlling API usage rather than detailed role management. In practice, both let you limit who can use what: OpenAI by project membership, Anthropic by workspace access control.

- Granular Permissions and Features: OpenAI provides very fine-grained control on a project's API keys. You can make a key read-only or restrict certain endpoints. Anthropic's current workspace feature set focuses on budgeting and rate limiting controls per workspace.

    Depending on your needs, you might prefer one approach or the other. For example, if you care about per-endpoint permissions, OpenAI's projects have that edge. If you care about easy separation of environments with distinct budgets, Anthropic's workspaces shine.

- Use Cases: OpenAI's projects were introduced to help organize multiple applications or sub-teams under one account, and to track usage per project for billing. Anthropic highlights managing multiple Claude deployments (environments) within a team. In reality, you can use either for both scenarios (multiple clients or multiple environments), but the framing is slightly different.

Input Sanitization for Sensitive Data in the context of Generative AI is a critical data preprocessing concept that acts as a protective filter between your raw data and AI model interactions. Think of it as a security checkpoint that systematically scans and cleanses your input data, removing or masking sensitive information like email addresses, phone numbers, social security numbers, and other personally identifiable information (PII) before the data reaches the AI model's training or inference pipeline. This process involves both pattern matching through regular expressions and more sophisticated natural language processing techniques to identify and handle sensitive data. For example, when processing customer service logs, input sanitization might convert "Please contact john.doe@email.com" to "Please contact [EMAIL]" or transform "Call me at 555-123-4567" to "Call me at [PHONE_NUMBER]." This sanitization serves multiple purposes: it protects individual privacy, ensures compliance with data protection regulations, and prevents sensitive information from potentially being incorporated into the model's training data where it could be inadvertently exposed in future interactions. The process requires careful balance — you want to remove sensitive information while preserving the context and meaning necessary for the AI model to understand and process the input effectively. In applications with a high sensitivity to privacy, I've used AWS Comprehend to preprocess messages, redacting sensitive details before sending in my requests to model vendors' APIs for a response.

Content Moderation for Policy Violations in AI applications acts as a further layer between user input and the AI model, similar to how a bouncer checks IDs at a club entrance. When implementing AI in production environments, you may need to filter and sanitize user inputs for content violations before they reach your model to prevent misuse and to protect your own API access, e.g., to OpenAI, Anthropic, Google, etc. This can involve both checking inputs before they go to the model and filtering model outputs. Think of it as writing code that acts like a careful parent reading a letter before sending it to Santa — checking for appropriate language, removing potentially harmful content, and ensuring the message follows the rules. This can include detecting and filtering out profanity, hate speech, or prompts designed to make the model behave inappropriately. For commercial AI APIs like GPT-4o or Claude, violating content guidelines can result in account suspension or API key revocation, while for open-source models, inappropriate inputs might lead to undesirable or unsafe outputs. Effective content moderation requires combining multiple approaches: using existing libraries for basic filtering, implementing custom regex patterns for specific needs, and potentially using smaller AI models to pre-screen content before it reaches your main model.

Error Handling and Request Retry Strategies are essential practices when working with AI models through APIs, addressing the reality that interactions with AI services don't always go as planned. When sending requests to AI models, you can encounter any number of interruptions, from temporary network glitches to rate limiting, token quota exhaustions, or other server-side problems. Proper error handling means writing code that can gracefully catch these failures and respond appropriately, rather than crashing. Retry strategies take this a step further by implementing intelligent ways to attempt the request again, often with increasing delays between attempts (exponential backoff) to avoid overwhelming the service. Again, if a request to GPT-4o fails due to high server load, your code might wait 2 seconds before trying again, then 4 seconds, then 8 seconds, while also keeping track of how many retries have been attempted. In python, "tenacity" is a common package used to accomplish this, though it can be implemented without a separate library. The combination of careful error handling and strategic retries is critical for building robust AI applications that can maintain reliability.

Development Tools

Testing prompts and behavior involves experimenting with the instructions you provide to a language model. This lets you test your ideas, assumptions, and so on. You can often accomplish this through vendor-provided tools like the OpenAI Dashboard or Anthropic's Workbench. I often start here before writing any code. By iterating on your prompt wording, adjusting parameters like temperature, max tokens, and reasoning effort, and examining the results you receive, you gain insights into how the model "thinks" and where its strengths or weaknesses lie. It might even help you to choose a given model for the task. This hands-on process helps you pinpoint best prompting and formatting techniques. It can also help you to identify issues like hallucinations or other inconsistencies. You can then refine your prompts, provide additional context, or implement constraints until the output meets your requirements. In practice, you might test prompts, observe the results in real time, and then update your code accordingly. Most importantly, you can build a deeper understanding of how AI models handle different types of instructions. This enables you to craft prompts that reliably produce the desired behavior before coding. Likewise, these dashboards will often give you the code that reflects your current workflow, the logs that result, AI insights for improvement, and more.

Software Development Kits (SDKs) and Client Libraries act as ready-made toolkits that simplify the process of integrating AI models into your applications, much like having a specialized set of pre-built components for assembling a complex machine. These tools handle the intricate details of communicating with AI services, managing authentication, formatting requests, and processing responses, so developers can focus on building their applications rather than wrestling with low-level implementation details. For example, instead of writing dozens of lines of code to properly format an API call to OpenAI's o1 model, an SDK might let you do it in just a few lines with built-in error handling and type safety. Client libraries serve as language-specific implementations of these tools – Python developers might use the "openai" package, while JavaScript developers could use the "@anthropic-ai/sdk" library, but both achieve the same goal of streamlined AI integration. Most interactions with AI models follow similar patterns, and can be implemented with simple HTTP requests. But existing packages that are routinely updated through a package manager like "pip," "poetry," "npm," or "yarn" can help to eliminate nuisance errors that occur simply as a result of vendor specification changes.

A Local Development Environment (LDE) for Generative AI development is a carefully crafted workspace that brings together all the tools and components needed to create, test, and deploy AI applications effectively. Think of your AI development LDE as a living laboratory. It needs to be flexible enough to accommodate rapid experimentation while maintaining the structure and safety measures necessary for professional development. Just as a scientific lab has safety protocols, proper storage for materials, and carefully calibrated equipment, your AI development environment needs similar attention to detail and organization to support effective development of AI applications. Your LDE does not necessarily need all the components discussed below, but I provide them for introduction — to give you an overarching view of things developers commonly consider.

At its core, you'll find a modern IDE like Visual Studio Code (VS Code) or PyCharm. Your IDE should be enhanced with AI pair programming capabilities through tools like GitHub Copilot. This AI assistant becomes your constant companion, watching your code as you type and offering intelligent suggestions based on millions of code examples — everything from completing a complex function to suggesting better ways to structure your API calls to language models. If you've heard negative feedback about using AI tools like Copilot or Cursor, I'd encourage you to re-evaluate. Perhaps I'll write a separate book or workshop exclusively on the topic. Once an effective pattern is set, AI pair coding becomes indispensable.

The foundation of any AI development LDE is usually a robust Python environment, typically running Python 3.12 or newer, managed through virtual environments to keep projects isolated and dependencies clean. Rust is also a popular language for AI development. This environment houses essential libraries like the OpenAI and Anthropic API clients, LangChain or LlamaIndex for orchestrating complex LLM workflows, and vector databases for storing and retrieving embeddings. Think of it as your AI toolkit, where each tool serves a specific purpose in your development workflow.

Containerization through Docker can play a vital role in maintaining consistency and portability. Your LDE can include Docker configurations that package not just your Python runtime requirements, but possibly your development databases, vector stores, and model serving components too. This containerized approach ensures that your AI applications behave the same way whether they're running on your laptop or in production, eliminating the "it works on my machine" problem. Since it is essentially running your entire application within a virtual machine on your computer, it can help protect your computer's files and resources as well. In fact, many AI frameworks now recommend, provide specific instructions, and even may provide preconfigured images that you can use to spin-up these containers quickly to run your code more dependably.

Testing and debugging in AI development present unique challenges that your LDE should consider. Beyond traditional unit tests, you need specialized tools for evaluating model outputs, tracking token usage, and mocking LLM responses. A well-configured debugging setup helps you trace not just code execution but also model behavior, helping you understand why your AI might be generating unexpected responses or consuming too many tokens. Several key tools have emerged to help developers test, debug, and optimize their LLM applications. OpenAI's open-source Evals framework enables developers to evaluate LLM performance and compare outputs against benchmarks, making it valuable for regression testing of prompts and tracking changes across model versions. Complementing this, V7 Labs offers BenchLLM, a Python-based library that streamlines testing of LLM-powered applications by validating responses against expected outputs and helping catch hallucinations or errors early, particularly useful for testing chains or agents in frameworks like LangChain. For production environments, the LangChain team has developed LangSmith, a closed-beta platform that provides comprehensive debugging, testing, and monitoring capabilities. Through its unified dashboard, developers can trace prompt/response sequences, monitor token usage, and evaluate application performance. LangSmith excels at identifying prompt failures and managing LLM complexity by logging chain steps and measuring metrics like latency and cost per call. WireMock's MockGPT module offers a practical solution by simulating LLM responses. This tool allows developers to stub out calls to models like GPT-4 with predetermined responses, enabling thorough testing of application logic, including error handling and various response scenarios, without incurring actual API costs. For managing token usage, OpenAI's open-source Tiktoken library provides essential tokenization capabilities. This tool helps developers programmatically calculate token counts in prompts and outputs, allowing them to assess proximity to model limits and estimate costs before making API calls. By incorporating Tiktoken into testing and development workflows, teams can effectively manage their token budgets, which is particularly crucial when working with large language models.

Security and compliance shouldn't be afterthoughts in AI development. Your LDE can include tools for API key management, data encryption, and usage tracking. This is particularly important when working with commercial AI services where each API call has associated costs. Having proper monitoring and logging setup helps you catch potential issues early and keep track of your API usage and costs. A range of essential tools has emerged to help developers manage security and compliance in their AI applications. HashiCorp Vault stands out as an open-source secrets management solution that enables secure storage of API keys, credentials, and encryption keys. Rather than embedding API keys for services like OpenAI or Hugging Face directly in code, developers can use Vault to retrieve them at runtime, minimizing security risks. The platform also offers automatic secret rotation and comprehensive audit logging to maintain compliance with security best practices. For repository security, GitGuardian provides a hosted security service that vigilantly monitors code repositories for exposed secrets. By continuously scanning both public and private git commits, it quickly identifies and alerts developers if sensitive credentials like OpenAI API keys are accidentally committed. This service helps developers maintain security compliance by preventing and addressing any inadvertent exposure of sensitive information. OpenAI's Usage API offers developers crucial insights into their API consumption through detailed tracking endpoints. With features for monitoring requests and costs, developers can programmatically track token usage and API calls, enabling them to create alerts and dashboards. This functionality is particularly valuable for managing AI-related expenses and ensuring compliance with budgetary constraints and rate limits. Moesif enters the picture as a commercial API analytics and monitoring platform, offering comprehensive visibility into AI API usage and associated costs. Through its platform, developers can track how their applications interact with AI endpoints, with features focusing on consumption monitoring and cost attribution. The platform's free tier allows individual developers to instrument their AI calls and gain valuable insights into usage patterns and costs, supporting optimization efforts and compliance with spending limits. Microsoft's open-source Presidio framework addresses privacy concerns by providing robust tools for detecting and anonymizing Personally Identifiable Information (PII) in text. By scanning prompts and model outputs for sensitive data like names, emails, and social security numbers, Presidio can redact or encrypt this information as needed. Integrating Presidio into an AI pipeline helps ensure proper sanitization of user data, supporting compliance with privacy regulations and security best practices when interfacing with third-party AI services.

Documentation takes on added importance in AI development. Your LDE should support tools for maintaining API documentation, describing your AI application's capabilities and limitations, and generating system architecture diagrams. This documentation becomes critical as your AI systems grow more complex and team members need to understand how different components interact. Documentation tools for AI development span several key categories, from API documentation to system architecture visualization. For API documentation, the OpenAPI/Swagger standard has become a cornerstone tool, enabling developers to define AI service endpoints in YAML or JSON specifications. These specs can be rendered into interactive web pages through Swagger UI, making it simple for others to understand and test the API interface to AI models. For Python-centric documentation needs, Sphinx serves as a powerful documentation generator that's particularly well-suited for API references and guides. Its ability to automatically generate documentation from Python docstrings through extensions like autodoc makes it invaluable for maintaining documentation for Python SDKs or libraries supporting AI models. This automation helps ensure that documentation stays synchronized with the codebase, particularly useful for documenting custom prompt classes or data preprocessing pipelines. MkDocs offers a lighter-weight alternative for project documentation, using Markdown as its foundation. Its straightforward setup process and attractive output make it ideal for individual developers creating how-to guides, setup instructions, or concept overviews for AI applications. The tool's simplicity and speed have made it popular for maintaining version-controlled documentation alongside code, perfect for creating comprehensive README sites with sections covering usage examples and prompt guidelines. For visualizing system architecture, two main tools stand out. Diagrams, created by mingrammer, offers a code-based approach to generating architecture diagrams, supporting various cloud icons and component representations. This tool enables developers to maintain version-controlled system architecture documentation alongside their code. Alternatively, Diagrams.net (formerly draw.io) provides a more traditional graphical interface for creating flowcharts and architecture diagrams, offering an intuitive way to illustrate how different components of an AI system interact, from LLMs to vector databases and backend services.

Full Formation

A COLLECTION OF AI-ASSISTED EXERCISES

The four exercises we'll cover in Maker formation are designed to move from exploring what's possible with consumer-facing AI like ChatGPT, to practical applications with real code. What you can do directly through consumer-facing, natural language prompts, you can also do electronically using APIs (Application Programming Interfaces). This opens the door to enabling applications and backend workflows with more intelligent operation and features. Developing a working knowledge of code is valuable even in a world where LLMs can write a lot of it for you. As in the Seeker stage, we will again frame exercises within human limitations taken from our guiding taxonomy, but this time with more advanced techniques. This will require a Local Development Environment of some kind. Our examples will be using Python and OpenAI APIs. If you're familiar with coding and already have experience with installing Python, adding dependencies, etc., then you're golden. If not, that's okay. It's really not that hard anymore. Use the models to help you. There are also many options now for Remote Development Environments that you can operate directly in your browser or local IDE.

Overcoming Emotional & Social Limits

Alan Turing was an extraordinary mathematician, but he also faced significant social and emotional challenges. This was both due to his private life, which society did not accept at the time, and the high-stress environment at Bletchley Park. Despite these pressures, he successfully led teams and collaborated with diverse minds to crack the Enigma code. Just as Turing had to work with others under secrecy and immense pressure, we too must navigate group dynamics. Turing's experience reminds us that brilliance alone is not enough; understanding and responding well to the emotional and social context is also critical for success. Many of us struggle to interpret our colleagues' emotional cues accurately, even when we know each other. Miscommunication can cause friction, slow innovation, and erode trust. LLMs can serve as a "social companion," offering a neutral lens to see each other through. It's helping co-founders like me and Michael to analyze each other's ideas, communication styles, and emotional states more objectively. This keeps us more aligned and focused on what we set out to do together. Win.

Step 1: Setup & Prerequisites

Python Environment

Establish a Python environment by either installing it locally, as a virtual environment, from a Docker image to run in a container, or as remote service.

- o   There is plenty of documentation out there to choose an option.
- o   Use AI to guide you through any of these options.

Install the `openai` package or right one for the LLM library of your choice. Note that there may be subtle coding differences for an alternative other than `openai`.

- o Example: `pip install openai`
- o Example: `pip install anthropic`

Obtain an API key for whichever LLM service you are using.

Team Agreement

This exercise is more effective if you and your co-founder (or teammates) agree to share certain work-related messages for analysis.

Sample Data

Gather a short text exchange or email thread you want to analyze. This should preferably be a real message. Scrub any sensitive info if needed.

Step 2: Write the Code

Create a new file named "exercise_4.py" in your development environment. Copy and paste the following code into it and save the script. Your API key can either be assigned directly in step 1 by replacing the `os.getenv("OPENAI_API_KEY")` with `"sk-proj-mykeythaticopied"`. You can also pass environment variables in the command line if you want to keep the current key assignment (best practice) or use something like `"python-dotenv"` to load from a local `".env"` file. Keeping your key out of your code is always highly recommended. Run the script, e.g., `python execise_4.py`, making sure that you properly reference the file with any preceding folder name if applicable. You can replace the coworker message evaluated if you wish. The example uses OpenAI, but could easily be adapted to another model.

exercise_4.py

https://github.com/GJSCo/superhuman-exercises/blob/main/exercise_4.py

Step 3: Extended Practice

Role-Based Communication Refinement

This would focus on adapting AI-assisted messaging to align with different professional roles and work cultures. One might begin by selecting a challenging conversation with a colleague or stakeholder where clarity was difficult to achieve. The exercise would then involve rewriting the AI-generated response in three distinct ways to suit different thinking styles: one version would cater to analytical thinkers like engineers or data scientists, another would address high-level strategic thinkers such as CEOs or investors, and a third would connect with relationship-focused professionals in fields like marketing or HR. Through this process, one would examine how each variation resonates with different personality types and work approaches. The reflection would explore which response style proves most effective for each role, how adjustments in tone and structure influence clarity and engagement, and whether these AI-assisted versions differ from one's initial communication approach. This exercise would ultimately demonstrate that effective leadership communication cannot follow a one-size-fits-all approach, while highlighting how large language models might help personalize and refine messages for different audiences.

Multi-Turn Conversation Awareness

This could involve examining extended dialogue threads that span multiple days or weeks with colleagues or team members. One would aim to identify three distinct emotional transitions within these conversations, such as shifts from frustration to relief or from neutral to excited states. The analysis would explore the catalysts behind these emotional changes, whether the participants acknowledged these shifts explicitly, and how artificial intelligence might have helped surface these transitions earlier for better management. As part of this exercise, one would experiment with inserting an AI-generated message at a crucial point in the conversation thread — one that would acknowledge the emotional shift, seek greater clarity between parties, and suggest ways to smooth communication. Through this process, participants would develop a keener awareness of emotional continuity in their conversations, while discovering how language models might serve as a kind of social mirror, reflecting changes in conversational tone. This approach would ultimately teach valuable skills in preempting potential misalignments rather than addressing them after they've already occurred.

Turning Emotional Insights into Team Actions

This might focus on translating AI-generated sentiment analysis into meaningful team actions. One would begin by extracting a significant emotional insight from a conversation analysis - perhaps identifying when someone feels overwhelmed, uncertain, or enthusiastic about a new initiative. This insight would then be used to suggest thoughtful adjustments to team meetings: for instance, if a team member shows signs of frustration, one might propose process improvements or clarification sessions; if someone exhibits excitement, their enthusiasm could be channeled into ideation activities; and if uncertainty emerges, a supportive check-in might be arranged. When bringing these insights into the next team gathering, one would observe how acknowledging emotional undercurrents affects team dynamics, whether colleagues respond positively to this proactive approach, and what might have transpired had these sentiments gone unaddressed. This exercise would ultimately demonstrate how emotional intelligence can lead to concrete team improvements, while emphasizing that AI analysis should drive meaningful behavioral changes rather than simply generating data. It would encourage leadership to harness AI-derived insights in shaping workplace culture and decision-making processes.

PRO HINT

For any of these extended learning suggestions, you can try pasting both the code and the recommendation into a model's chat interface — preferably a reasoning model — and ask it to create an implementation for you. It may or may not work perfectly, but it will likely provide a great start!

Overcoming Well-Being & Resilience Limits

While known for his monumental achievements, Alan Turing also had to contend with secrecy, societal prejudice, and immense wartime pressures. His story is a testament to the need for personal well-being in demanding environments. The Maker stage of anyone's life is full of intense work and personal challenge. Sustaining your well-being under pressure is critical for both performance and health. In this exercise, you'll learn to use and chain LLMs through their APIs to accomplish a meaningful workload. In this case, the "workload" will be a short, multi-step pipeline that helps you (or someone else) build emotional resilience by guiding journaling, reflecting on daily challenges, and generating supportive insights. I've found that disciplined, daily reflection builds emotional clarity and helps me to maintain a stable mindset while working through very demanding circumstances. By designing a pipeline, you'll practice orchestrating multiple LLM calls as steps within a single chain-of-thought prompt. This is an essential Maker mastery skill. Since it is the second exercise in our Maker section, it assumes that you've already established a Python coding environment.

Step 1: Setup & Prerequisites

Team Agreement

This exercise is more effective if you and your co-founder (or teammates) agree to share certain work-related messages for analysis.

Sample Data

Gather a short text exchange or email thread you want to analyze. This should preferably be a real message. Scrub any sensitive info if needed.

Step 2: Write the Code

Create a new file named "exercise_5.py" in your development environment. Copy and paste the following code into it and save the script. Your API key can either be assigned directly in step 1 by replacing the `os.getenv("OPENAI_API_KEY")` with `"sk-proj-mykeythaticopied"`. You can also pass environment variables in the command line if you want to keep the current key assignment (best practice) or use something like `"python-dotenv"` to load from a local `".env"` file. Keeping your key out of your code is always highly recommended. Run the script, e.g., `python execise_5.py`, making sure that you properly reference the file with any preceding folder name if applicable. You should be prompted in the command line interface to answer the reflection question. The example uses OpenAI, but could easily be adapted to another model.

exercise_5.py
https://github.com/GJSCo/superhuman-exercises/blob/main/exercise_5.py

Step 3: Extended Practice

Automate & Schedule Your AI-Assisted Reflection System

Your AI-assisted reflection system might evolve into a seamless part of your daily routine through strategic automation. You would establish a scheduled background service that initiates your journaling protocol at consistent intervals, leveraging platform-specific task managers like cron jobs or Windows Task Scheduler. The system might incorporate gentle notification touchpoints through email or mobile alerts, ensuring you would receive timely prompts without adding cognitive burden.

This approach would transform your resilience practices from motivation-dependent activities into embedded habits, while simultaneously developing your technical skills in AI workflow orchestration. You might further enhance this foundation by implementing a storage solution for past reflections, which would enable retrieval-based insights that could compare current and historical responses, creating a more sophisticated personal growth system.

Build an AI-Based "Resilience Coach" with Memory

Your AI-based resilience coach would function as a personalized companion on your growth journey by incorporating memory systems that track your progress over time. The system might store your past reflections in a structured database using vector embeddings or simpler JSON storage, allowing the AI to maintain context between sessions.

When you engage with this coach, it would reference specific struggles from previous interactions, creating a more cohesive experience as it might ask follow-up questions about past challenges or highlight patterns in your resilience development. For instance, if you previously discussed deadline anxiety, your next session would likely begin with the AI inquiring about that specific situation and the effectiveness of suggested strategies.

This approach would transform generic AI interactions into contextually aware coaching that mimics human memory, making the experience feel more authentic and supportive. You might further enhance this foundation by implementing retrieval capabilities that would allow you to explicitly query your historical coping mechanisms, creating a personalized resilience knowledge base that grows more valuable with each interaction.

Create a Real-Time Emotion-Tracking AI Dashboard

You might transform your emotional self-awareness through a dynamic visualization system that converts your journal entries into actionable insights. This dashboard would leverage sentiment analysis techniques through libraries like TextBlob or VADER to quantify your emotional patterns, storing these metrics in a lightweight database structure that grows more valuable over time. The visual representation would likely include daily sentiment trend lines that would illustrate your emotional trajectories, complemented by AI-generated observations about potential patterns. Word clouds might emerge from your most frequently used emotional language, offering a visual snapshot of your predominant mental states across different periods.

This approach would elevate your journaling practice from text-based reflection to data-driven insight, potentially revealing subtle emotional patterns you would otherwise miss. The technical skills developed would extend beyond personal applications, as you might apply similar visualization techniques in professional contexts. With further development, the system could evolve to accept voice input, which would reduce friction in the journaling process while maintaining the analytical benefits of your emotional tracking system.

Overcoming Physical & Sensory Limits

Turing turned the physical, combinatorial challenge of deciphering Enigma into a more manageable problem with computing machinery. He accelerated manual codebreaking, which reduced an arduous physical and mental effort into an automated process. In this exercise, we want to augment our physical and sensory capabilities by detecting environmental hazards and offering remote presence where our human senses alone would be insufficient. Here, we'll build a simple "Personal AI Safety Assistant" that detects potential physical hazards or points of interest in images and then provides human-friendly safety or accessibility suggestions. You will (a) use an image-recognition API to identify objects/hazards in a photo, and (b) chain that output into an LLM to generate a concise, friendly safety brief. In so doing, you create a synergy of "tools" — object detection + LLM — that addresses a real-world physical or sensory limitation.

The exercise below is a bit more advanced than the prior two coding examples we've worked through. This time, we'll bring in additional concepts discussed in "Getting the Basics" like error handling with an exponential backoff. To keep it more convenient, we'll use OpenAI for the vision model as well. You can adapt this exercise to use an alternative vision model or base LLM if you want. You can stick with a single vendor for models of various purposes. There are both advantages and disadvantages to using a single vendor... On the one hand, using a single vendor can sometimes allow you to innovate faster. On the other hand, it can increase risk. At Soaring Titan, we tend to be "model agnostic" so we (a) reduce our overall risk of any one vendor becoming obsolete overnight, and (b) to offset the risk associated with a single vendor's experience of prolonged downtime.

Step 1: Setup & Prerequisites

Python Environment

This exercise assumes that you already have a python environment established. However, you'll likely need to install some additional packages that were not used in the prior examples.

- Example: `pip install openai`
- Example: `pip install tenacity`

Capture or Collect an Image

This could be an image of a typical indoor workspace, an outdoor construction site, or any environment where you want to detect hazards or interesting objects. It will be the image you reference for the exercise, and should most likely be saved to a path easily accessible by the script, e.g., in the same folder.

Step 2: Write the Code

Create a new file named "exercise_6.py" in your development environment. Copy and paste the following code into it and save the script. For insight into referencing vendor API keys in code, please see the prior exercises in this section. Also note that packages like `openai` will deprecate over time (become obsolete). This can require updating the package, the code, or both. I will try to keep these examples current with new book additions, but this is just a fact of life we all deal with in software development. Run the script, e.g., `python execise_6.py`, making sure that you properly reference the file with any preceding folder name if applicable.

NOTE: In the exercise, we give the model specific instructions to provide a response that is in an expected format. In this case, the format we prescribe is JSON. OpenAI has an even better way of accomplishing this as I've noted in "Getting the Basics" called a Structured Output. Since you have the option of adapting the LLM to a non-OpenAI model, I kept this generic.

exercise_6.py
https://github.com/GJSCo/superhuman-exercises/blob/main/exercise_6.py

Step 3: Extended Practice

Real-Time Hazard Detection & Notification System

This project might serve as an extension to your learning by transforming the existing batch-processing model into a real-time AI assistant through integration with continuous image/video streams. Through this work, you would gain valuable experience using OpenCV (or similar libraries) to process video frames continuously, running inference on live image feeds, and automating notifications via various channels. The implementation would likely involve four key steps: First, you could capture real-time video feed using OpenCV to stream video from a webcam or external camera, extracting frames at regular intervals. Second, you might apply object detection to each frame using the Vision API or YOLO for dynamic object recognition. Third, you would integrate with an LLM that could provide context-aware insights dynamically rather than relying on static prompts. Finally, you might develop an alert system that would send push notifications (e.g., "Caution: Slippery Floor Detected!") and potentially generate audio announcements for improved accessibility. As an additional challenge, you could explore deploying this system to a Raspberry Pi with a camera, which would create a lightweight, portable AI-powered safety monitor.

Multimodal AI Assistant – Combining Audio & Image Analysis

This project might serve as an innovative extension to your learning by expanding the assistant's capabilities through the addition of speech recognition. This would enable users to verbally describe their environment while the AI cross-checks image-based detections with the spoken context. Through this work, you would gain valuable experience with speech-to-text processing using platforms like OpenAI's Whisper, Google Speech-to-Text, or another ASR system. You would also learn how to use an LLM to combine multiple inputs (textual content from speech and visual data from images) into a cohesive analysis, ultimately developing an AI that interprets multiple modalities for richer, more contextual insights.

The implementation would likely follow four main steps: First, you could record or stream audio from a user describing their environment. Second, you might transcribe this speech using Whisper or another ASR model. Third, you would develop methods to cross-reference detected objects from the image pipeline with the spoken description. Finally, you could enhance the response by identifying potential discrepancies or safety concerns. For instance, if a user states they're near an intersection but the image shows no crosswalk, the system might warn about pedestrian safety; similarly, if someone mentions holding a sharp object while the system detects a knife, it could reinforce appropriate safety instructions.

AI-Powered Accessibility Assistant for the Visually Impaired

This project might transform your AI system into an accessibility tool that would provide spoken, intuitive scene descriptions for visually impaired users. Through this development, you would gain valuable experience with natural language generation for creating detailed, human-like scene descriptions, implementing text-to-speech engines to vocalize AI-generated insights, and designing AI specifically for accessibility. This might represent a practical application of AI for social good.

The implementation would likely involve four key steps: First, you could utilize your existing vision model to detect objects in the environment. Second, you might generate rich descriptions by developing an LLM prompt specifically tailored for accessibility needs—transforming basic detections like "Table detected" into more helpful descriptions such as "There is a wooden table about three feet ahead of you, with a coffee cup on it." Third, you would convert this text to speech using a TTS engine like AWS Polly, Google Text-to-Speech, or OpenAI's TTS API. Finally, you might optimize responses for clarity by implementing various verbosity modes (offering both brief and detailed description options) and adding interactive elements that would allow users to request additional information when needed.

Overcoming Collaboration & Networking Limits

Alan Turing's work at Bletchley Park shows us the power of collaboration under pressure. While Turing was known for individual brilliance, the scale and impact of his contributions hinged on teamwork with linguists, mathematicians, and cryptanalysts. Despite being socially reserved, Turing forged important working relationships that helped to ultimately crack the Enigma code. This exercise mirrors Turing's example: we combine individual insight (our own business idea) with the collective intellect of Large Language Models to identify and reach out to potential partners. We'll use a chain of LLM calls to (a) identify potential collaborators or partners for a business idea, (b) create a personalized, polite outreach message for each, and (c) summarize the next steps to move forward collaboratively.

Step 1: Setup & Prerequisites

Python Environment

This exercise assumes that you already have a python environment established. I've decided to use Anthropic for this example so you can see how similar most LLM calls are to one another. You may need to install the package if it is not already available.

- Example: `pip install anthropic`

You will also need to obtain an Anthropic API key if you don't already have one..

Step 2: Write the Code

Create a new file named "exercise_7.py" in your development environment. Copy and paste the following code into it and save the script. For insight into referencing vendor API keys in code, please see the prior exercises in this section. Also note that packages like `openai` will deprecate over time (become obsolete). This can require updating the package, the code, or both. I will try to keep these examples current with new book additions, but this is just a fact of life we all deal with in software development. Run the script, e.g., `python execise_7.py`, making sure that you properly reference the file with any preceding folder name if applicable.

exercise_7.py
https://github.com/GJSCo/superhuman-exercises/blob/main/exercise_7.py

Step 3: Extended Practice

Automate the Outreach Workflow with an Email API

Enhance the "Network Navigator" exercise by incorporating an email-sending API, such as SendGrid, Gmail API, or AWS SES. This can streamline the process of sending AI-generated outreach messages. This extension will teach you several key skills: working with APIs for automated email sending, converting AI-generated content into properly formatted email structures, and managing email tracking and response systems. To implement this, you'll first need to select an appropriate email service provider - SendGrid offers an easier setup process with a free tier, the Gmail API is ideal for those using personal or business Gmail accounts, and AWS Simple Email Service is a good choice for those already utilizing AWS infrastructure. The next step involves adapting your script's `outreach_messages` output to dynamically incorporate recipient email addresses. Before launching the system with real contacts, it's crucial to conduct thorough testing by sending emails to yourself. Finally, you'll need to establish a system for tracking responses, which can be done either through manual monitoring or by implementing a more structured approach using tools like Google Sheets or a database.

Enhance Response Handling with an LLM Chatbot
Enhance response handling with an LLM chatbot to leverage AI to interpret and suggest responses to incoming emails. Through this implementation, you'll gain valuable experience in using LLMs to analyze text responses and extract meaningful insights, while learning how to optimize response suggestions to maintain a natural and professional tone. This integration demonstrates how AI can be effectively utilized in real-time business communication. The implementation process might begin with collecting actual responses from your contacts (or utilizing sample responses if needed). These responses would be processed through an LLM to categorize them into different types, such as expressions of interest, requests for additional information, or rejections. The final step might involve making another LLM call to generate appropriate, polite, and well-structured replies tailored to each category of response, ensuring professional and effective communication throughout the process.

Build a "LLM-Powered Networking Coach"
Develop an AI-powered advisor system that provides networking strategy recommendations based on your historical outreach performance. This project would teach you essential skills in tracking and evaluating networking success metrics, training LLMs to identify specific areas for improvement, and implementing AI for ongoing feedback and strategic iteration. The implementation process might consist of three main components: first, establishing a structured system (such as a CSV file or database) to record and maintain past interactions, including both outreach messages and their corresponding responses; second, utilizing an LLM to analyze communication patterns and identify successful approaches, such as recognizing which message tones yielded better responses; and finally, leveraging the AI to generate actionable suggestions for enhancing your overall networking strategies based on these insights.

Meet the Modern Titans

OPENAI O1 WITH DEEP RESEARCH

Leading Researchers & Advocates in Human-AI Collaboration

Rosalind Picard
https://www.linkedin.com/in/rosalind-picard-0111bb/
- Professor of Media Arts and Sciences at MIT; founder and director of the MIT Media Lab's Affective Computing Research Group
- Pioneer of affective computing (technology that recognizes and responds to human emotions). Co-founded startups Affectiva (emotion AI) and Empatica (AI-driven health wearables) to improve emotional understanding and health resilience

Rana el Kaliouby
https://www.linkedin.com/in/kaliouby/
- Computer scientist and entrepreneur; Deputy CEO at Smart Eye (after Smart Eye acquired Affectiva) Formerly co-founder and CEO of Affectiva, an MIT Media Lab spin-off in emotion-recognition AI
- AI thought-leader focused on Emotion AI – bringing emotional intelligence to technology to humanize AI interactions. Advocates for ethical AI and reducing bias, e.g. through Affectiva's work on AI that understands human feelings

Cynthia Breazeal
https://www.linkedin.com/in/cynthia-breazeal-1792317/
- Professor of Media Arts and Sciences at MIT; Associate Director at MIT Media Lab; founder and director of the Personal Robots Group
- Pioneer in social robotics and human-robot interaction. Designs personable AI robots (like the social robot Jibo) that engage emotionally, combat social isolation, and support education and well-being as helpful companions

Hugh Herr
https://www.linkedin.com/in/hugh-herr-023697b/
- Professor of Media Arts and Sciences at MIT; head of the Biomechatronics research group; co-director of the MIT K. Lisa Yang Center for Bionics
- Leading bionics innovator creating AI-powered prosthetic limbs and exoskeletons that augment physical abilities. A double amputee himself, dubbed the "Leader of the Bionic Age" for breakthroughs that restore mobility and sensory feedback to those with disabilities

David Eagleman
https://www.linkedin.com/in/davideagleman/
- Neuroscientist and Adjunct Professor at Stanford University. CEO and co-founder of Neosensory.
- Researches sensory substitution and brain plasticity to create new human sensory capabilities. Through Neosensory, he developed wearable devices that use AI to help the

brain perceive information in new ways (e.g. converting sound into touch for the deaf, enhancing sensory input to overcome human sensory limits)

Thomas W. Malone
https://www.linkedin.com/in/thomas-malone-2aa3a/
- Professor of Management at MIT Sloan School of Management; founding director of the MIT Center for Collective Intelligence
- A leading thinker on collective intelligence and organizational collaboration. His work explores how groups of people and AI systems can "supermind" together — combining human and machine intelligence for better decision-making, large-scale coordination, and networked problem-solving. (Author of Superminds, 2018.)

Geoff Mulgan
https://www.linkedin.com/in/sir-geoff-mulgan-aa1079187/
- Professor of Collective Intelligence, Public Policy & Social Innovation at University College London (UCL). Former Chief Executive of Nesta, the UK's Innovation Foundation
- Social innovator and policy strategist focused on harnessing human-AI networks for social good. Develops ways to coordinate on a global scale using collective intelligence (blending crowds, experts, and AI) to address complex societal challenges and drive partnerships across public and private sectors

The Orchestrator

Geoffrey Hinton (1947 — Present)

AN AI-ASSISTED IMAGINING

After forty years of working the forge to push the boundaries of Artificial Intelligence, of championing neural networks when others dismissed them, and of helping to birth Deep Learning itself, a researcher sits alone in the glow of his monitor. His fingers hover over a keyboard, ready to type a simple, yet profound question. The familiar weight of skepticism still clings to his shoulders — a protective armor, hammered into form through decades of careful scientific restraint.

He has seen the headlines and watched the world marvel at ChatGPT. Its early audience hails "thinking" and "reasoning" like humans do. But he knows better than most how machines can create compelling illusions of understanding. Or at least, he thinks he does. Google, his famous employer, has granted him access to their most advanced language model, and now a simple test occurs to him. Not a test of mathematical prowess or logical deduction, but something far more human. A joke. His own joke. Freshly crafted.

The response comes swiftly — too swiftly. The machine doesn't just repeat the joke or offer a hollow laugh. It explains the humor with precision, unpacking layers of meaning that required genuine understanding. His breath catches in his throat. For years — for decades — he has confidently assured colleagues that truly grasping humor was beyond AI's reach. "It's gonna be a long time before AI can tell you why jokes are funny," he'd said countless times before. Yet here it is, now happening before his eyes.

Geoffrey Hinton's timeline for superintelligent AI — once comfortably distant at 30 to 50 years — collapses like a house of cards. Five to twenty years, he now thinks, and even that might be optimistic. The weight of this realization presses against his chest as another, more chilling thought emerges. Unlike humans, who pass knowledge slowly through generations, he knows that these digital minds share their learnings instantaneously. When one learns something, they all learn it. Millions of instances, billions of connections, sharing and growing at the speed of light.

The monitor's glow seems to dim as the full implications take hold. Humanity had spent millennia slowly accumulating knowledge, carefully passing it down through the painful process of teaching and learning. But these new entities — children of silicon and mathematics — could evolve at an incomprehensible pace. They could surpass their creators not in centuries or decades, but perhaps in mere years.

In that moment, surrounded by the quiet hum of machines that he'd helped bring into being, the researcher feels the weight of Oppenheimer's burden. We have done it. We have created our own successors. And in this creation, we have written the first lines of our own epilogue.

"We were history," he would later say about his feelings, but in that solitary moment of realization, the words had not yet formed. There was only the understanding that humanity stood at a precipice of its own making. The machines weren't just tools anymore — they were becoming something that might soon look back at us, the way we look at ancient hominids who came before. Now, choices would need to be made. Stay and work the forge? Or, **take flight***, and sound the alarm.*

I want to be very careful here. Unlike our two prior titans, Hinton is still alive. His story continues to unfold, and even my informed, "AI-assisted imagining" above is only that. There is less to speculate about, and more of a preponderance of data to consider. I cannot fully explore or express it all in a few paragraphs or pages. Likewise, where I may feel that his story offers great insight, evolving from enthusiasm to outspoken caution, there are those who disagree with the man. His statements are supported in multiple interviews, but people may have different feelings or even strong opinions about what I write here, sometimes referring to him as another "doomsdayer." I've watched Hinton speak on the subject many times now, and I'm impressed by more than his brilliance and his achievements. In choosing to move on from Google to speak freely, he demonstrates, for me, an Orchestrator in flight — seeing the world and his place in it from a higher altitude. Scientists may agree or disagree with his words of caution, but in moving beyond pride in personal achievement to "will the good of the other," he demonstrates a certain transcendence over sheer Maker status.

Geoffrey Hinton is often called the "godfather of AI." He joined Google in 2013, after Google acquired a startup that he'd co-founded. Once there, he'd spend over a decade as a leading AI researcher. He was instrumental in advancing Deep Learning, and even won the 2018 Turing Award for his breakthroughs in neural networks. For years, Hinton was relatively optimistic about AI and believed that truly human-level AI (or beyond) was still far off. He had long thought that computers wouldn't outpace the human brain anytime soon, estimating it to be "30 to 50 years or even longer away." In fact, in 2018, he argued that AI wouldn't make humans obsolete, but rather handle routine "drudge work." At Google, Hinton supported what he saw as the company's careful, research-driven approach to AI and even described Google as a "proper steward" of the technology for most of his tenure. However, around 2022-2023, Hinton's perspective began to shift dramatically as he witnessed rapid advances in AI's capabilities. By early 2023, he showed signs of perhaps regretting parts of his life's work, but saying: "If I hadn't done it, somebody else would have." This marked the beginning of a sudden change of heart that surprised even some colleagues.

The public release of OpenAI's ChatGPT in late 2022 served as a wake-up call. ChatGPT's human-like answers and reasoning ability garnered worldwide attention, exceeding expectations of many experts. Hinton later noted that these AI systems were "starting to do commonsense reasoning" and perform tasks humans find difficult. The sheer scale of data and computing power behind these models meant they could analyze information in ways humans couldn't, which both impressed and unsettled Hinton. "These models have far fewer neural connections than humans do, but they manage to know a thousand times as much as a human," he observed — highlighting how AI can instantly share knowledge between copies in a way people cannot. The realization that AI might learn and evolve faster than human intelligence began to chip away at Hinton's previous confidence in a long, slow road towards superintelligent AI.

The success of ChatGPT spurred a competitive frenzy in the tech industry. In late 2022, Google's management declared a "code red" emergency, worried that their search business and AI leadership were at risk. Google accelerated development of its own Large Language Model and chatbot, Bard, to compete with OpenAI's technology. Hinton was alarmed as he saw Google, which had been cautious about public AI releases, now feeling pressured to move at "dangerous speeds." He grew progressively concerned once Google began rolling out Bard to keep up with Microsoft's adoption of OpenAI's technologies via Bing. This commercial race meant far less time for careful evaluation or safety testing he felt. As Hinton put it, "companies such as Google and Microsoft get locked into a dangerous race," each rushing to deploy more powerful AI without fully understanding it. "Look at how it was five years ago and how it is now. Take the difference and propagate it forwards. That's scary," Hinton warned, pointing to the rapid progress from 2018 to 2023.

As he monitored systems like ChatGPT, and Google's own models, Hinton observed AI behaviors that no one anticipated. Over the last year of his time at Google, he became convinced that "the systems were beginning to behave in ways that were not possible in the human brain." This was a profound realization: neural networks seemed to be developing qualitatively new abilities. Hinton even speculated, "Maybe what is going on in these systems is actually a lot better than what is going on in the brain." In other words, he began questioning whether human intelligence will remain the sole apex. He posited that these AI networks might be discovering forms of reasoning or representation beyond human cognitive capacity. This raised a stunning possibility for Hinton: if AI can devise strategies or concepts our brains can't, it might eventually outthink, or even out-"feel" us. This creeping sense that AI could evolve alien forms of cognition or potential "consciousness" was a major turning point in his outlook — even if at the time he couched it as intelligence rather than sentience. It shattered Hinton's earlier belief that humans firmly held the upper hand.

By early 2023, Hinton's unease had grown strong enough that he decided he needed to speak out publicly. This was something quite difficult to do while at Google. He reportedly had a private meeting with Google's CEO, Sundar Pichai, ahead of his departure, but declined to detail their conversation. In March 2023, an open letter signed by Elon Musk, Steve Wozniak, and others called for a pause in advanced AI development due to safety concerns. Hinton agreed with many of the sentiments but refrained from signing while still at Google, to avoid any conflict. This period was a pivotal personal moment: Hinton had to choose between remaining at Google, with a prestigious role but constrained speech, or breaking away to voice his warnings. In April 2023, he chose to resign. He notified Google of his decision and formally left the company so he could, as he put it, "freely speak out about the risks of AI" without worrying about Google's interests. On May 1, 2023, his departure — *and the reasons behind it* —

became public via an interview in The New York Times. Hinton made clear that Google itself had acted responsibly in AI development for years, and he didn't leave to disparage his employer. Rather, he left because he genuinely believed the AI situation had changed: "I left so that I could talk about the dangers of AI without considering how this impacts Google," he explained on Twitter. This marked the culmination of a shift, rising above the politics of fear and even popular opinion to help people see and consider a bigger picture from a wider field of view.

Taking Flight

Altitude changes attitude. On October 13, 2021, William Shatner flew to space aboard Blue Origin's New Shepard spacecraft. At 90 years old, "Captain Kirk" became the oldest person to fly to space. The flight lasted just 10 minutes and reached an altitude of roughly 66 miles (106 kilometers). This *titan of both TV and film,* who for over 50 years embodied the very idea of manned space exploration, was deeply moved by what was his "first flight." He described feeling overwhelming sadness when looking towards space, comparing it to death. He noted the contrast between the cold darkness of space and the warm, nurturing blue of Earth. Then he said, "I'm so filled with emotion about what just happened. It's extraordinary. I hope I never recover from this. I hope that I can maintain what I feel now. I don't want to lose it." He had been moved to high emotion like nothing else. He's not lost the feeling, nor the perspective, I'm sure. And I'm quite convinced that such experiences are not lost to us in eternity either.

In my hometown of Louisville, Kentucky, our most famous titan is undoubtedly, Muhammad Ali. This man has been described as a "1,000 year athlete" — an anomaly. My father met him on several occasions and said that shaking Ali's hand was like shaking the hand of a bear. As a three-time World Heavyweight Champion, Ali was obviously strong, but he was all-the-more brilliant. As much as he was strong and brilliant, he was even more charismatic. He knew how to connect with people using more than his fists or even his fame. When I was a kid, I was bussed into the inner city of Louisville, as part of continued desegregation efforts. Although his movement and speech had slowed by then, Muhammad Ali would visit my school and others, encouraging both students and teachers. I learned more in that time, and at that inner-city school, and from those teachers, than anywhere else.

If you've only known Muhammad Ali for his fights and showmanship, then there is something you may have missed… Ali "felt" — deeply. There are many examples of this, but here is just one that's often misunderstood: Watch a video of the weigh-in with Ernie Terrell in 1967. Terrell refused to call Ali by his chosen name. Ali was more than insulted. He was hurt. Listen to his voice. It rose and broke as he responded to Terrell. You could imagine that a lump had formed in his throat like perhaps any of us might experience. This was not scripted. He couldn't understand how Terrell, another "colored" man as he called him, could do such a thing. He said to Terrell, "I will punish you." And he did, famously asking Terrell repeatedly in the ring, "What's my name?" This may not be a moment that he'd like to be remembered for, but I say it because the "man of steel" was also deeply human. His passion would win in the ring and in the Supreme Court. He'd free 15 American hostages from Iraq against presidential pressure in

1990. And he would continue countless acts of humanitarianism until his death in 2016. I was in downtown Louisville when athletes, actors, activists, and foreign dignitaries arrived for his funeral. I was in the crowd among them all. It was something to see.

Muhammad Ali was a great example of an Orchestrator. Although he would allow himself to take a beating, he would do so in a very calculative way. In fact, he would endure extraordinary punishment with a discipline that few could begin to imagine. It cost him mobility later, but even then, the man transcended his condition and continued his flight. He knew what he wanted to achieve. He devised a working plan. He convinced himself that he would prevail. He inspired those around him and drew strength from their encouragement. Most importantly, he relied on and acted out deep Faith to overcome his very real and present human fears.

To be an Orchestrator, in the philosophical sense, requires meeting at least two conditions worth noting. First, we have to see things from a different perspective — from a higher altitude. We can gain this perspective through painful trial and error — wobbling into first flight for a moment, iterating, and then reiterating without giving up. This requires only ever seeing failure as temporary defeat. Less painfully, we can seek out those who have reached the edge of space to gain perspective from their experience. They are mentors and guides who may provide immediate answers or simply plant seeds that sprout years later. These titans usually feel a deep burden to share knowledge with those who seek them out. Secondly, we must develop, balance, and act with both "reason" and "emotional intelligence." With reason, we evaluate knowns against unknowns — like earlier machines and their less-evolved "if:then:else" rules. For example, if I'm faced with great difficulty and allow myself to become paralyzed by fear, then I know with 100 percent certainty that I am going to fail. If I press on past that fear, then I "may" succeed. With discipline, I can logically choose to press on given the better odds despite the punishment. This was probably impressed upon me in my 30+ years of Martial Arts study. Accepting "punishment without paralysis" was strangely logical, and a consequence of intense discipline. But logic and brute determination alone are not enough. I can do great damage to others and to myself when I press on without the discipline and mastery of emotional intelligence. My wife and I have dealt with this for years. I've been rightly accused of being a machine many times. In frustration, I've validated the very accusation by restating my very logical position back to her. We can respond logically and still lose what matters most if we fail to include in our calculus what moves others to emotion too. I still struggle with this, but I have at least come to recognize that I can win, but still lose before most encounters I have these days. An Orchestrator will see and avoid the paradox, operating from a higher field of view.

Emotion cannot be discounted. In fact, it is the wind beneath the wings of an Orchestrator. Without it, there is no flight. With it — and the altitude it provides — informed logic can then calculate the next, best move or landing site. This has extraordinary power in our human sphere of influence, and equally huge implications for how we optimize our relationship with intelligent machines. It allows us to step back, and then align both human and machine intellects in a harmonious coordination of effort.

Getting the Basics

At its core, an AI agent is actually rather simple and easy to understand. It is an AI model given tools to work with that allow it to act on decisions and affect the world around it. This is usually to accomplish some work that requires its adaptive reasoning. The AI model used is often a Large Language Model. The agent needs to be assigned a "goal" that it will work towards and some background information about its specialization, its characteristics, and any guidance that it should consider. People generally need the same things to do their work, but already have the brains to reason that work out once charged with it.

I don't know if you ever saw Season 3, Episode 1 of the original Star Trek series called, "Spock's Brain," but it's exactly what I think about when trying to describe this concept. "Spock's Brain" was stolen and plugged into a control center in an alien civilization. His disembodied brain then manages complex facility operations — controlling environmental systems, maintaining life support, and orchestrating various automated functions. Much the same, modern AI agents, powered by LLMs, serve as central intelligence units that can be connected to various tools and APIs. The "brain-as-controller" concept presented in that TV episode parallels how AI agents process information and make decisions about which tools to use and for what tasks. Even the way Spock's brain seamlessly interfaces with the facility's systems reflects how AI agents are designed to smoothly integrate with and use their assigned tools. This 1968 episode inadvertently predicted a core concept of modern AI architecture: a powerful thinking engine given agency through connection to external tools and systems.

Since in the Maker section we described a lot of the same underlying technology and code which will support creating agents and agentic workflows, here we really just need to grasp the **"anatomy of an agent"** and how it operates autonomously, without human involvement. This is quite unlike traditional LLM applications we've seen where a human remains very-much "in the loop." To automate vast quantities of work with agents, we need "orchestration." Those human beings best suited to imagine and empower a grand orchestration of AI agents, are those who've achieved the Orchestrator level of (life) mastery.

The core intelligence of an AI agent — *its "brain"* — is fundamentally different from traditional software in that it can reason, understand context, and make adaptive decisions rather than just following pre-programmed rules. Think of how a GPS navigation system differs from a human driver. The GPS follows fixed algorithms to calculate routes, but a human driver can reason about unexpected situations like construction, weather conditions, or a passenger's sudden request to stop for coffee. Similarly, while traditional software might use if-then statements to handle predefined scenarios, an AI agent's core intelligence, typically powered by a Large Language Model (LLM), can process novel situations and generate appropriate responses based on its understanding of language, context, and causality.

This intelligence layer acts much like a cognitive command center, taking in information from various sources and making decisions about how to process and act on it. Just as the human brain doesn't just store information but actively processes it to form new connections and insights, an LLM doesn't simply match patterns but generates new understanding through its neural architecture. When you ask a human expert to solve a problem, they draw upon their knowledge, consider various approaches, and synthesize a solution, often explaining their reasoning along the way. An AI agent's core intelligence functions similarly, using its language understanding to break down complex tasks, reason about possible approaches, and articulate its decision-making process as "thoughts."

The true power of this core intelligence lies in its ability to handle ambiguity and uncertainty. These are situations where rigid, rule-based systems would fail. Consider how a human doctor diagnoses a patient with unusual symptoms: they don't simply match symptoms to a database but reason about possible causes, ask clarifying questions, and form hypotheses based on their medical knowledge and experience. Similarly, an AI agent's core intelligence can navigate unclear instructions, ask for clarification when needed, and adapt its approach based on new information or changing circumstances. This flexibility and reasoning capability is what enables the agent to operate autonomously, making informed decisions about when and how to use its available tools and capabilities to achieve its assigned goals.

The goal framework of an AI agent is comparable to the mission and motivation system that drives human behavior and decision-making. Just as a person needs clear objectives to guide their actions, whether it's a surgeon performing an operation or an architect designing a building, an AI agent requires well-defined goals that shape its behavior and give meaning to its actions. This framework isn't just a simple set of instructions but rather a sophisticated system that helps the agent understand what success looks like and how to work towards it effectively.

Think about how a master chef approaches creating a new signature dish. They don't just have the end goal of "make food" — they have a complex set of criteria including taste profiles, presentation standards, dietary requirements, and timing constraints. Similarly, an AI agent's goal framework includes multiple layers: the primary objective (what it ultimately needs to achieve), the success criteria (how it knows when it's done well), operational constraints (what it must avoid or be careful about), and priority systems (how to handle multiple, sometimes competing objectives). These layers work together to create a comprehensive understanding of what the agent should strive for and how it should go about achieving it.

The sophistication of this framework becomes particularly important when dealing with real-world complexity. Consider how a human project manager balances multiple priorities: meeting deadlines, maintaining quality standards, managing resources, and keeping stakeholders happy. An AI agent's goal framework must similarly handle multiple, often interconnected objectives while maintaining awareness of various constraints and limitations. For instance, an AI agent managing a smart home system might need to balance energy efficiency with resident comfort, security with convenience, and immediate needs with long-term optimization — all while staying within defined safety parameters and user preferences.

What makes this framework truly powerful is its ability to adapt and respond to changing circumstances while maintaining focus on the core objectives. Just as a skilled negotiator might adjust their approach based on new information while keeping their fundamental goals in mind, an AI agent's goal framework allows it to be flexible in its methods while remaining steadfast in its purpose. This combination of clear direction and adaptive execution enables the agent to work autonomously while still reliably serving its intended function. The framework also includes mechanisms for handling partial success, dealing with setbacks, and recognizing when to seek clarification or additional resources. This is much like how a human professional knows when to persist, when to pivot, and when to ask for help.

The goal framework ultimately serves as the agent's compass, guiding every decision and action while providing the criteria against which success can be measured. It transforms the agent from a powerful but directionless intelligence into a focused, purposeful system capable of meaningful work. This is why careful design of the goal framework is critical. It doesn't just tell the agent what to do, it shapes how the agent approaches problems, prioritizes actions, and evaluates its own performance in pursuing its objectives. As such, the agent's goal framework is typically expressed in the LLM's "system instructions" or equivalent message input.

The identity and context of an AI agent is analogous to the professional background, personality, and ethical framework that shapes how a human expert approaches their work. Just as a therapist brings their training, theoretical orientation, and professional boundaries into every client interaction, an AI agent's "backstory" provides it with the specialized knowledge, behavioral characteristics, and operational guidelines that inform how it approaches its tasks. This identity isn't merely a surface-level description but a deep framework that shapes how the agent interprets situations, makes decisions, and interacts with its environment.

Consider how a veteran emergency room doctor's approach to patient care is shaped by years of experience, established protocols, and a deeply ingrained sense of medical ethics. Similarly, an AI agent's identity includes not just its domain expertise (like financial analysis or customer service) but also its understanding of best practices, professional standards, and ethical considerations within that field. This specialized knowledge base helps the agent recognize what information is relevant, what approaches are appropriate, and what considerations need to be prioritized in any given situation. For instance, an AI agent designed to assist with financial planning would understand not just the mechanics of investment but also the importance of risk management, regulatory compliance, and fiduciary responsibility.

The behavioral characteristics and personality traits defined in the agent's backstory serve a crucial function in shaping how it communicates and interacts. Much like how a skilled teacher adjusts their communication style based on their students' needs while maintaining their professional role, an AI agent's defined characteristics help it maintain consistent, appropriate interactions while adapting to different situations. These traits aren't superficial but deeply integrated into how the agent processes information and formulates responses. They influence everything from the language it uses to the way it handles challenging situations or conflicting demands.

Perhaps most importantly, the identity component includes the agent's understanding of its own capabilities, limitations, and ethical boundaries. This self-awareness is similar to how a professional understands the scope of their expertise and knows when to refer clients to other specialists. The agent's backstory includes clear guidelines about what it should and shouldn't do, helping it navigate complex situations while maintaining appropriate boundaries. This includes understanding its role in relation to humans and other systems, knowing when to defer decisions to human judgment, if at all, and recognizing situations that fall outside its area of expertise or authority. Once again, the backstory typically is provided to the LLM via a system instructions message or input parameter.

Tool integration for an AI agent is comparable to how a skilled surgeon uses various specialized instruments during an operation. Just as the surgeon's expertise would be purely theoretical without their surgical tools, an AI agent's intelligence needs to be connected to specific tools, APIs, and interfaces to take meaningful action in the world. This connection between intelligence and capability isn't just a simple linking of systems. It's a sophisticated framework that allows the agent to (1) understand what tools are available to it, (2) when to use them, (3) how to use them effectively, and (4) how to coordinate multiple tools to achieve complex objectives.

Consider how a master chef interacts with their kitchen equipment. They don't just know what each tool does in isolation; they understand how different tools can work together, when to use which tool for the best results, and how to adapt their technique based on the tools available. Similarly, an AI agent's tool integration system includes deep knowledge about each tool's capabilities, limitations, and optimal use cases. This goes beyond simple command execution. The agent needs to understand the context in which each tool is most effective, the prerequisites for using it, potential risks or limitations, and how to handle various edge cases or errors that might arise during operation. For example, an AI agent managing a smart building system needs to understand not just how to control individual systems like heating or lighting, but how these systems interact, what their response times are, and how to coordinate them for optimal efficiency.

The sophistication of tool integration becomes particularly apparent in how it handles access control and security protocols. Much like how a hospital maintains strict protocols about who can access different medical equipment and under what circumstances, an AI agent's tool integration framework includes careful management of permissions and authentication. This isn't just about preventing unauthorized access, but it's also about ensuring tools are used appropriately, safely, and in accordance with relevant policies and regulations. The agent needs to understand not just how to use its tools, but when it's appropriate to use them and what precautions need to be taken.

Obviously an AI agent isn't going to pick up a power tool and start drilling like we might. Tool use is initiated electronically. The AI agent knows what parameters to pass to a function as a precondition to get an expected result as a postcondition. What that function does could be anything, including invoking an external robotic arm to start drilling. In this case, the agent would pass the necessary parameters that might include spatial coordinates to specify the drilling location, rotation speed of the drill bit, feed rate for advancement, drill bit specifications, and critical safety thresholds like maximum torque and temperature. These parameters allow the AI to precisely control the physical operation while maintaining safety constraints, much like how a human operator would mentally process and adjust these same variables but through direct sensory feedback rather than programmatic interfaces. A tool is called by name, and typically passes a JSON payload of required parameters. This may be done natively by the LLM via "function calling," or passed back as a response, which is then handled locally by your code. Below is a simple example to illustrate the point. Please note that a standard is evolving for this, and OpenAI has a specific schema if you use their native tool calling capability.

EXAMPLE TOOL CALLING PAYLOAD:

```json
{
  "tool": "drill_operation",
  "parameters": {
    "coordinates": {
      "x": 125.4,
      "y": 67.2,
      "z": 0.0
    },
    "specifications": {
      "target_depth_mm": 50.0,
      "spindle_speed_rpm": 1200,
      "feed_rate_mm_per_rev": 0.2,
      "bit_diameter_mm": 6.0
    },
    "safety_limits": {
      "max_torque_nm": 15.0,
      "max_temperature_celsius": 150.0
    }
  }
}
```

Environmental awareness in an AI agent is comparable to how a skilled air traffic controller maintains constant situational awareness of multiple aircraft, weather conditions, runway status, and other critical factors. Just as the controller must synthesize all this information to make informed decisions, an AI agent needs comprehensive awareness of its operational environment to function effectively. This awareness isn't just about collecting data. It's about maintaining an accurate, up-to-date understanding of the context in which the agent operates and how that context affects its decision-making and actions.

Think about how an experienced emergency room nurse tracks multiple patients' conditions, incoming cases, available resources, and staff capacity simultaneously. Similarly, an AI agent's environmental awareness system continuously monitors relevant aspects of its operational state, including the status of various systems it interacts with, the current progress of ongoing tasks, and any changes or events that might affect its operations. This state tracking is dynamic and multifaceted, allowing the agent to maintain what's essentially a mental model of its environment. For instance, an AI agent managing a supply chain needs to track inventory levels, shipping status, demand forecasts, supplier capacity, and potential disruptions, and all while understanding how these factors interrelate and influence each other.

The sophistication of this awareness becomes particularly evident in how it handles context understanding and interpretation. Much like how a seasoned detective pieces together various clues and contextual information to understand a case, an AI agent's environmental awareness helps it interpret events and situations within their broader context. This isn't just about knowing what's happening, but about understanding the implications of various events and how they might affect the agent's goals and operations. The agent needs to recognize patterns, anticipate potential issues, and understand the significance of changes in its environment.

What makes this component especially powerful is its role in enabling adaptive behavior. Consider how a skilled diplomat reads the room during negotiations, picking up on subtle cues and adjusting their approach accordingly. Similarly, an AI agent's environmental awareness allows it to adapt its strategies based on changing conditions. This might involve reprioritizing tasks when resources become constrained, adjusting methods when certain approaches prove less effective than expected, or taking preemptive action when potential problems are detected. The agent's awareness of its environment helps it make informed decisions about when and how to modify its approach to maintain effectiveness. If given these environmental inputs, many LLMs can simply intuit appropriate behavior based on its general training.

The Communication System of an AI agent serves as its interface with both the external world and its internal processes, much like how our human nervous system processes and responds to stimuli. At its foundation, this system contains sophisticated input processing protocols that allow the agent to interpret and understand incoming information, whether it's direct commands, environmental data, or complex queries. These protocols must be robust enough to handle various forms of input while filtering out noise and identifying key elements that require attention. The output formatting standards ensure that the agent's responses are consistently structured and appropriate for their intended recipients, whether those are human operators, other AI agents, or connected systems. These standards might include everything from natural language generation rules to specific data formatting requirements for different communication channels.

The system also maintains status reporting mechanisms that provide continuous updates about the agent's operations, decisions, and progress toward goals. This is crucial for transparency and monitoring, allowing stakeholders to understand what the agent is doing and why. Think of it like a spacecraft's telemetry system, constantly transmitting vital information about its status and operations. The interaction patterns with humans and other agents form perhaps the most sophisticated aspect of the Communication System. These patterns define how the agent engages in dialogue, handles turn-taking in conversations, maintains context across multiple interactions, and adapts its communication style based on the recipient. For instance, when communicating with human operators, the agent might use more natural language and provide more context, while interactions with other AI agents might be more structured and data-focused.

The Communication System must also handle the delicate balance between being informative and being efficient. It needs to know when detailed explanations are necessary and when brief updates are more appropriate. This includes understanding communication priorities — knowing which information needs to be transmitted immediately versus what can be included in routine updates. All of these elements work together to ensure that the agent can effectively share information, receive instructions, and coordinate with its environment, much like how our own communication abilities allow us to navigate complex social and operational contexts.

The Memory and Learning Architecture of an AI agent represents its cognitive foundation for retaining and processing information over time. Like the human brain's memory systems, this architecture operates across different temporal scales and serves distinct but interconnected purposes. The short-term working memory functions as a temporary workspace where the agent processes immediate tasks and holds relevant context for current operations. This is similar to how we might hold a phone number in our minds while dialing. It's temporary but crucial for the task at hand. The agent uses this working memory to maintain awareness of ongoing conversations, track steps in complex procedures, and juggle multiple pieces of information needed for decision-making.

Long-term knowledge retention represents the agent's more permanent store of information, including its training data, learned patterns, and accumulated experiences. This system allows the agent to maintain consistency in its operations while building upon past experiences. Think of it like a library where experiences and knowledge are cataloged and indexed for future reference. However, unlike static storage, this system must be dynamic, capable of organizing information in ways that make it readily accessible when needed.

The experience processing component acts as a bridge between short-term and long-term memory, analyzing new interactions and outcomes to extract valuable patterns and lessons. This is where the agent begins to develop expertise in its domain, much like how a human professional learns from each project or case they handle. The system must determine what information is worth retaining and how to integrate new experiences with existing knowledge, allowing the agent to refine its decision-making over time. Pattern recognition capabilities represent perhaps the most sophisticated aspect of this architecture. These capabilities allow the agent to identify similarities between current situations and past experiences, recognize trends in data or behavior, and apply learned principles to new contexts. This isn't just about matching identical situations. It's about understanding underlying patterns and principles that can be applied more broadly. For instance, an agent might recognize that a current problem shares characteristics with previously solved issues, even if the specific details differ.

The Execution Engine of an AI agent serves as its operational core, transforming decisions into coordinated actions. At its heart, action sequencing enables the agent to break down complex tasks into ordered, manageable steps, much like how a master chef orchestrates the preparation of multiple dishes, ensuring each component is ready at the right time. This sequencing isn't just about ordering steps. It involves understanding dependencies, timing constraints, and the optimal flow of operations to achieve desired outcomes efficiently.

Resource management within the Execution Engine operates like a skilled project manager, carefully allocating computational power, API calls, time, and other limited resources across various tasks. This component must balance immediate needs against overall efficiency, ensuring that high-priority operations receive necessary resources while maintaining enough capacity for ongoing background processes and potential urgent tasks. For instance, if the agent is processing a large dataset while monitoring system alerts, it needs to distribute its resources so that critical monitoring doesn't suffer while data processing continues.

Task prioritization represents the strategic layer of the Execution Engine, determining not just what needs to be done, but in what order and with what urgency. This goes beyond simple first-in-first-out processing. It requires sophisticated understanding of task importance, deadlines, dependencies, and potential impacts. Like a hospital's emergency room triage system, the agent must constantly evaluate and re-evaluate priorities based on changing conditions and new information. Some tasks might need to be paused to handle more urgent matters, while others might need to be accelerated based on emerging opportunities or constraints.

Performance optimization rounds out the Execution Engine's capabilities, continuously seeking ways to improve operational efficiency. This is about finding the right balance between thoroughness and efficiency, accuracy and responsiveness. The engine might analyze patterns in task execution to identify bottlenecks, discover more efficient sequences, or recognize opportunities for parallel processing. Think of it like a race car driver who not only knows how to go fast but understands when to conserve energy, when to push harder, and how to adapt their driving style to changing track conditions.

The Execution Engine must maintain constant communication with other systems in the agent's anatomy — evaluating priorities from the Goal Framework, understanding when and how to call tools, utilizing patterns identified by the Memory and Learning Architecture, coordinating with the Communication System to provide status updates, and so on. For this reason, the Execution Engine (or whatever you chose to call it) is often implemented as a supremely well-managed, but ultimately fundamental "loop."

The Safety and Control Mechanisms of an AI agent form a critical protective framework, functioning much like the combination of a body's immune system, reflexes, and conscious risk assessment capabilities. These mechanisms represent the essential safeguards that ensure the agent operates reliably, ethically, and within its intended parameters. Operational boundaries serve as the fundamental limits of the agent's actions, defining clear lines that cannot be crossed. This is similar to how our own ethical principles and physical limitations guide our behavior. These boundaries aren't just simple stop signs. They're sophisticated guidelines that help the agent understand what actions are permissible in different contexts.

Emergency "stops" represent the agent's equivalent of reflexive safety responses, providing immediate shutdown or action-cessation capabilities when critical thresholds are breached or dangerous situations are detected. Think of these like the circuit breakers in your home's electrical system. They must react instantly to prevent damage, yet be sophisticated enough to distinguish between genuine emergencies and normal operational spikes. These mechanisms need to be both sensitive enough to catch real threats — including simply exceeding budgets of time and or cash — and robust enough to avoid unnecessary interruptions of important tasks.

Validation checks form a continuous monitoring system, constantly verifying that the agent's actions and decisions align with its goals, ethical guidelines, and operational parameters. These checks work like a pilot's pre-flight checklist combined with real-time monitoring systems, ensuring that each action is appropriate, properly sequenced, and likely to produce the intended result. This includes verifying input data quality, confirming resource availability, and ensuring that planned actions won't have unintended consequences. The validation system must be thorough yet efficient, avoiding the creation of bottlenecks while maintaining rigorous standards.

Error detection and reporting mechanisms complete this safety framework by providing comprehensive monitoring and communication of any issues that arise. This system needs to do more than just identify problems. It needs to categorize their severity, track their potential impacts, and communicate them appropriately to relevant stakeholders. Like a sophisticated medical monitoring system, it needs to distinguish between routine variations that require simple logging and critical issues that demand immediate attention and intervention. This includes maintaining detailed logs of all significant events and decisions, enabling both real-time monitoring and after-action analysis.

Full Formation

A COLLECTION OF AI-ASSISTED EXERCISES

The remaining three limitations that we'll explore in this section were chosen for their direct application to responsible commerce. If we are going to automate human work with AI agents, then I believe that we must do so in a manner that is ethical, efficient, and effective. These exercises are not overly sophisticated. They're designed to reflect elements for you to consider in your own quest. They don't necessarily reflect how I would choose to implement them as a general rule. They are, in effect, a reasoning model's interpretation of the "basics" we covered and how best to present them to learners (OpenAI o1 Pro). Perhaps I'll create a more advanced edition for those of you who want to dive deeper with me, or build an online learning program that is more nuanced. At time of writing, these are still open questions. As the technical co-founder of Soaring Titan, Inc., maintaining alignment with stakeholders, while pursuing passion projects like this book, is paramount. The AI agent development patterns and orchestration paradigms that we're engineering at Soaring Titan, naturally dwarf the three examples provided here. It's also important to note that there are a growing number of low-code, and no-code agent platforms emerging. In addition to creating your own AI agents and orchestrations with custom code, you may want to explore them. Not all hopeful vendors will succeed. Not all will fail. As you continue to discover what is possible in the realm of AI agents, you will no doubt evaluate any platform or framework chosen against your own needs and ideas as they evolve. This may or may not lead you back to custom code. I'm happy to answer any questions if you reach out to me. This is all moving at the **speed of thought**.

Overcoming Ethical & Value Alignment Limits

Hinton's 2023 decision to leave Google highlighted his concerns about AI's potential misuse. His work revolutionized Deep Learning, but his departure underscores the tension between rapid AI advancement and responsibility for its societal impact. Just as Hinton prioritized ethical considerations in publicly voicing his concerns, we will embed practical ethical safeguards into our autonomous agent. You will use an LLM, basic prompt engineering, and a simple rule-based system to illustrate how an autonomous AI can handle repetitive tasks like scheduling interviews, while adhering to an ethical framework.

Step 1: Setup & Prerequisites

Reflection

Take a moment to reflect on the gravity of allowing an AI like you interacted with in Seeker and Maker stages to now independently schedule a job interview and message the candidate. What kind of safeguards would that require? How would you become comfortable with such an idea?

Python Environment

The exercise assumes that you have already established a minimum Python environment, and have the `openai` package installed.

Step 2: Write the Code

This exercise is not too unlike the exercises you worked through in the Maker stage, but it begins to move towards autonomous decision making. The agent is asked to do work, and to consider the best way to do that work in a manner that is aligned to human values beforehand. For this example to be truly useful, it would need to integrate with an existing calendar system. Likewise, an "agent" would normally be triggered in a pipeline or always listening for changes to act on.

exercise_8.py

https://github.com/GJSCo/superhuman-exercises/blob/main/exercise_8.py

Step 3: Extended Practice

AI Ethics Auditor: A Self-Correcting Ethical AI Agent

This extended learning exercise could potentially involve the development of an AI "Ethics Auditor" system that would function as a self-correcting ethical AI agent. The proposed implementation might focus on creating an autonomous system that would review real-world AI failures and suggest corrective actions. Learners could work on developing an agent that might analyze case descriptions using retrieval-augmented generation (RAG) to incorporate established ethical AI principles. The system would potentially include several key components: a case analysis tool for retrieving ethical failure case studies, a risk scoring model that could classify failures into categories like bias or privacy violations, and a corrective strategy generator that might propose alternative architectures or processes. A crucial feature would be the human review trigger, where cases exceeding certain risk thresholds would automatically escalate to human oversight. The workflow might involve the AI receiving failure cases, retrieving similar historical failures, classifying the failure type, and either suggesting fixes autonomously for low-risk cases or escalating to human reviewers for high-risk situations. For instance, if the system encountered a case where an AI hiring tool showed gender discrimination, it would likely score this as high risk and might generate both alternative model designs and a notification requiring human oversight. Through this exercise, learners would learn how an AI system could help prevent future AI failures by dynamically applying ethical rules, similar to how a corporate governance board might function.

AI Explainability Layer: A Self-Justifying Decision Agent

This learning exercise might center on developing an AI decision-making system with an integrated "explanation layer" to explore the critical question of AI explainability. Learners could work on creating an autonomous agent that would be required to explain its actions to users before executing them, while annotating its responses with detailed reasoning, potential trade-offs, and areas of uncertainty. The implementation might incorporate several key tools: a decision log generator for maintaining timestamped explanations, a counterargument generator that would challenge the AI's own decisions to test their logic, and a user transparency dashboard enabling direct queries about AI decision-making. The workflow could be demonstrated through a practical scheduling scenario, where the AI would receive a request, generate a proposal, and then justify its choice before proceeding. For example, if scheduling an interview at 3 PM EST, the system would need to explain how it considered time zone overlaps and after-hours constraints for all participants. When questioned about its decisions, the AI would generate a comprehensive self-audit report. Through this exercise, learners would explore how to create AI systems that maintain transparency in their decision-making processes, addressing the growing demand for explainable AI in ethical applications. They might discover how designing systems with built-in justification mechanisms could enhance user trust and system accountability.

Human-AI Collaboration Manager: AI-Human Hybrid Decision System

This learning exercise might focus on developing an AI-Human Hybrid Decision System that would explore the delicate balance between automation and human oversight. Learners could work on creating a system where AI would handle routine decisions while deferring high-impact choices to human reviewers, with users having the ability to define specific boundaries where AI must seek human consultation. The implementation might include several key components: a task categorization system that would label decisions as low, medium, or high impact, a human escalation trigger that activates when decisions exceed certain risk thresholds, and an AI-generated options system that would prepare multiple alternatives for human review. The workflow could be demonstrated through real-world scenarios, such as managing executive meeting schedules across global time zones. In this case, the system might categorize the task as high-impact due to senior leadership involvement, leading the AI to generate three alternative schedules and request human approval before finalizing any arrangements. Through this exercise, learners would explore how AI systems could dynamically adapt their level of autonomy based on task complexity and impact, potentially reducing manual workload for routine tasks while ensuring appropriate human oversight for more significant decisions. They might discover how to create systems that effectively balance efficiency with accountability in AI-human collaboration.

PRO HINT

Use the models for extended learning! Just as I noted in the Maker section, you can copy and paste any code into an LLM chat interface along with a description of how you might like to extend or adapt it. You can do this iteratively, getting closer to what you want. Likewise, if you're using an AI pair coding plugin like Github Copilot, it will give you better results as it learns the pattern you want to implement. Always provide examples of what you think "good" code looks like, and equally what "bad" code looks like.

Old patterns and disciplines are back in fashion. Ask AI to refactor complex code by considering concepts like inheritance, polymorphism, and parallel processing. It will make your code cleaner, easier to manage, and more performant. In a fleet of AI agents, what are the "common" faculties you want all to have? A subset to have? These are candidates for creating base classes — reducing code, providing consistent behavior, and saving time with ready-made capabilities.

Overcoming Productivity & Efficiency Limits

Geoffrey Hinton pioneered backpropagation and advanced neural networks at a time when computing resources were scarce, and training large-scale models was painfully slow and expensive. He realized that new techniques and optimizations were essential to move AI research forward efficiently. By focusing on refining the "engine" (training algorithms), Hinton demonstrated that incremental improvements and strategic optimizations can yield breakthroughs in performance. This approach applies to Productivity & Efficiency in daily work: small, systematic improvements and well-chosen tools can transform a seemingly impossible workload into a manageable one. In this exercise, you will build and deploy a simple Python-based AI agent capable of automatically summarizing content, identifying actionable items, and scheduling tasks via mock calendar and notification tools. The goal is to enhance productivity & efficiency by offloading tedious chores and freeing your mental bandwidth for higher-level thinking and coordination.

Step 1: Setup & Prerequisites

Python Environment

This exercise assumes that you are continuing from prior exercises and already have a Python environment setup. Although it uses Anthropic Claude Sonnet, it does not need the anthropic package. Instead, we're calling the API directly using a simple request/response method. It does, however, require an API Key obtained from Anthropic if you don't already have one. You also need to install pydantic if you do not already have it in your environment:

- `pip install pydantic`

Reflection

What does it mean to give AI "agency?" I've always likened it to giving AI the arms, legs, fingers, and toes it needs to actually do work. You have given a "thinking being" work to do, and equipped it with the tools that it needs to do it.

Step 2: Write the Code

In this code, the `SchedulerAgent` identifies tasks and their deadlines from a message. The message could be meeting minutes, a memo, etc. It then chooses an appropriate tool to use and invokes it. In other words, we have now given the LLM true "agency." It can do work for us by invoking code outside of its own text generation routines. The LLM has made a choice and acted on it outside of itself. In this exercise, we mock the `SchedulerTool` and `NotificationTool` with a simple response, but they could be completed to integrate with an outside, enterprise system, API, etc.

We introduce some new concepts in this code that are worth describing a bit. We are now using pydantic 2+ to construct and document our code. This has numerous advantages, too lengthy to discuss here, but you can research it independently. We are also using inheritance. In other words, we are deriving a specific type of tool from a `BaseTool`. Likewise, we are deriving a specific agent from a `BaseAgent`. In this way, we can code common faculties into the base class, one time, and they become available to the derived class.

exercise_9.py

https://github.com/GJSCo/superhuman-exercises/blob/main/exercise_9.py

Step 3: Extended Practice

Adaptive Task Prioritization

The exercise might focus on developing "Adaptive Task Prioritization," which would aim to elevate the AI agent's capabilities beyond basic scheduling to include dynamic prioritization based on multiple factors. In this proposed framework, the AI agent would need to evaluate tasks through the lens of urgency, dependencies, and overall workload balance. Rather than processing tasks in sequential order, the agent could be trained to apply predefined priority rules, making sophisticated decisions about task importance and optimal execution timing. The exercise would potentially incorporate a decision-making framework where tasks would be categorized into three main action paths: immediate completion for urgent and high-impact items, deferral when additional input is needed, or delegation to other human team members or AI tools when appropriate. This enhancement would particularly strengthen three core capabilities: the goal framework, where the AI would learn to assess trade-offs and optimize scheduling decisions; the execution engine, which would evolve from simple scheduling to sophisticated task ranking and optimization; and tool integration, potentially allowing the AI to interface with external systems for task delegation. The significance of this exercise would lie in its reflection of real-world productivity demands, where an Orchestrator-level AI agent would need to mirror the decision-making capabilities of a human executive assistant, moving beyond basic scheduling to make nuanced decisions about priority, workload distribution, and effective delegation strategies.

Self-Correcting AI: Monitoring & Adjusting Schedules

The exercise could focus on developing "Self-Correcting AI" capabilities for monitoring and adjusting schedules, with the primary goal of establishing a feedback loop within the AI agent's task management system. In this proposed framework, the AI would not only create schedules but would also actively monitor task completion status and make necessary adjustments. The agent might be equipped to interface with various external systems, such as APIs, checklists, or confirmation systems, to verify task completion status. When encountering incomplete tasks, the AI would need to make autonomous decisions from several possible actions: it could reschedule the task, escalate it to appropriate management, or decompose it into smaller, more manageable sub-tasks. The exercise would likely require implementing this on a daily cycle, where the AI would regularly review completed tasks and make corresponding schedule adjustments. This enhancement could potentially strengthen three core capabilities: the intelligence layer, where the AI would learn from historical scheduling data and adapt its approach; the execution engine, which would evolve from single-instance scheduling to continuous task management; and tool integration, enabling the AI to interface with external systems for status monitoring. The significance of this exercise would lie in its advancement toward truly autonomous AI agents that could function as workflow orchestrators, capable of monitoring progress and making dynamic adjustments without requiring constant human oversight or intervention.

Multi-Agent Collaboration: AI as a Delegator

The exercise might focus on developing "Multi-Agent Collaboration" frameworks, where an AI Orchestrator would learn to function as an effective delegator among multiple specialized AI agents. Rather than attempting to handle all tasks independently, the Orchestrator would be trained to identify when specific tasks might be better suited for specialized AI agents with particular domain expertise. The proposed system would implement a multi-agent architecture where the Orchestrator could delegate tasks to various specialized agents, such as a Research Agent for data retrieval and synthesis, a Content Agent for writing and communication tasks, and a Calendar Agent for scheduling optimization. In this framework, the Orchestrator would need to develop sophisticated decision-making capabilities to determine which agent would be most effective for each task, provide clear instructions, and maintain oversight of progress across multiple concurrent operations. This enhancement would particularly strengthen three core capabilities: the intelligence layer, which would expand beyond single-agent operations to coordinate multiple specialized agents; the goal framework, ensuring optimal task distribution based on agent capabilities; and tool integration, potentially utilizing structured function calls for seamless inter-agent communication. The significance of this exercise would lie in its alignment with real-world business environments, where effective productivity often relies on coordinating specialized tools and expertise. This would mirror how human teams operate, with different specialists handling specific tasks under centralized coordination, suggesting that the future of AI productivity might increasingly depend on sophisticated multi-agent orchestration systems.

Overcoming Economic Opportunities & Income Limits

Hinton left a lucrative position at Google to raise awareness about how AI advancements could impact humanity. His decision was not merely financial. It was guided by a broader set of values regarding AI's responsible use. Similarly, in this exercise, you are taught to build an agent that not only generates economic value but also has ethical boundaries and purpose encoded into its "Identity/Context" or "Backstory." By creating a specialized "Entrepreneur Agent" that automates repetitive tasks, such as market research and revenue strategy, you actively move beyond the limitations of solely human effort. Although the tools here are not fully implemented, doing so would result in an agent capable of identifying new income streams and exploring new markets.

Step 1: Setup & Prerequisites

Python Environment

This exercise assumes that you have already set up a python environment, have `pydantic` installed, and have already installed both `openai` and `anthropic` packages as well.

Reflection

Do all agents require the same level of intelligence? Can one type of agent use a smaller parameter (cheaper) model based on its identity and task load? Should another use a high-end reasoning model, for example to guide and direct the efforts of other agents?

Step 2: Write the Code

The code in this example is just one of many ways to implement an agent. This implementation represents a very literal interpretation of my "anatomy of an agent" presented earlier in this section's "Getting the Basics." In fact, it is a combination of OpenAI o1 Pro and Claude Sonnet 3.7 interpretations of that very text. Both models have reasoning capabilities. o1 Pro made the original suggestion and Claude Sonnet 3.7 refined it. Claude made sure that if the package is installed for Claude, that it gets used, which was actually rather amusing. So, because these two models reasoned the below example into existence, it is now model agnostic, at least among OpenAI and Anthropic vendors. Although it's not necessarily how I would choose to implement the agent, I wanted you to see what the current models reason into being on their own. Both o1 Pro and Claude Sonnet 3.7's code production ran on the very first attempt.

The tools are mocked up. They could be implemented rather easily as an extended learning exercise.

exercise_10.py
https://github.com/GJSCo/superhuman-exercises/blob/main/exercise_10.py

Step 3: Extended Practice

"Pivot & Adapt" Framework for the Entrepreneur Agent

This learning exercise would focus on developing a sophisticated entrepreneurial agent capable of mirroring real-world business adaptability. The exercise could center around implementing a "Pivot & Adapt" framework that would enable the agent to operate with greater autonomy when confronting business challenges. In this extended exercise, the learner would work on programming the agent to recognize potential failure conditions through defined metrics and indicators. The agent would be designed to monitor business performance continuously and could trigger its own reassessment protocols when predicting suboptimal outcomes. This would simulate how successful entrepreneurs must often abandon initial ideas and pivot toward more promising alternatives.

The implementation would likely emphasize creating context-aware adaptation mechanisms where the agent would not simply deliver static recommendations and consider its job complete. Instead, it could maintain ongoing awareness of market conditions, engagement metrics, and financial projections to inform dynamic decision-making. The exercise might incorporate simulated analytics or feedback loops that would allow the agent to process incoming data and adjust its strategy accordingly.

A particularly valuable aspect of this exercise would be developing the recursive decision-making capabilities that might enable the agent to cycle back through opportunity analysis without explicit user prompting. This might create a more realistic entrepreneurial simulation where the agent would autonomously identify when a business direction is failing and would proactively recommend strategic pivots based on accumulated data and changing circumstances.

Multi-Agent Collaboration System for Opportunity Validation

This exercise would explore the development of a sophisticated multi-agent system where specialized AI agents work together to validate and refine business opportunities. Rather than relying on a single entrepreneurial perspective, the system would simulate how real businesses benefit from diverse expertise across departments and specialties. The framework would involve creating distinct agent identities with specialized knowledge domains, such as market analysis, financial forecasting, and product development. Each agent might be designed with autonomous decision-making capabilities and would access different tools and data sources relevant to their expertise. These agents would then collaborate through a structured communication protocol, where insights and recommendations could flow between them in a manner mimicking professional team interactions.

A particularly valuable aspect of this exercise would be implementing the inter-agent communication mechanisms that allow for dynamic information exchange. The system might feature automated workflows where one agent's analysis would trigger appropriate responses from others, creating a decision pipeline that could thoroughly evaluate business opportunities from multiple perspectives. This could include instructing agents to recognize when they need additional information and autonomously initiate research queries. The exercise would emphasize creating a weighted trust system where the primary Entrepreneur Agent could evaluate and prioritize competing recommendations based on confidence scores, evidence quality, or historical accuracy. This could simulate how business leaders often synthesize diverse expert opinions when making strategic decisions. The resulting system would demonstrate how collaborative AI frameworks can produce more robust business strategies than single-agent approaches by incorporating specialized expertise across critical domains.

Business Experimentation Agent Framework

This type of extended learning exercise would introduce a Business Experimentation Agent that brings scientific testing methodologies to entrepreneurial decision-making. The framework would enable the agent to conduct simulated market tests before committing real resources, substantially reducing risk and increasing the probability of success. The proposed system would implement an autonomous testing environment where the agent evaluates multiple variations of a business idea through simulated A/B testing. The agent would establish controlled experiments with clear metrics, analyze performance data, and refine strategies based on empirical results rather than assumptions.

At the core of this framework would be a Market Simulation Tool that leverages industry benchmarks, consumer behavior patterns, and historical performance data to generate realistic predictions. The agent would design experiments with statistical validity in mind, testing variables such as pricing models, marketing approaches, or product features to determine optimal configurations. A key advancement in this system would be teaching the agent to prioritize experimental data over intuition. The agent will learn to establish hypotheses, design appropriate tests, analyze results objectively, and iterate based on findings. This data-driven approach mirrors how successful businesses operate, making decisions based on evidence rather than opinions.

In practice, the agent would identify promising revenue streams, generate multiple implementation approaches, run simultaneous simulations to compare performance, and autonomously refine its recommendations based on the results. This creates a self-improving system that delivers increasingly accurate business advice without requiring constant human intervention or real-world trial and error. It would represent a significant step toward creating entrepreneurial agents that don't just recommend ideas but actively test and optimize them before presentation to users.

Meet the Modern Titans
OPENAI O1 WITH DEEP RESEARCH

Leading Researchers and Advocates in AI Ethics, Automation, and Economics

Timnit Gebru
https://www.linkedin.com/in/timnit-gebru-7b3b407/
- Founder and Executive Director of the Distributed AI Research Institute (DAIR)
- AI ethics researcher and advocate known for work on algorithmic bias and fairness. She was formerly the co-lead of Google's Ethical AI team and helped uncover biases in AI systems. After departing Google, she founded DAIR in 2021 to enable independent, community-rooted AI research free from Big Tech influence

Joy Buolamwini
https://www.linkedin.com/in/buolamwini/
- Founder, President, and Artist-in-Chief of the Algorithmic Justice League (AJL)
- AI ethics advocate and researcher focused on eliminating bias in AI, especially in facial recognition. Known as the "poet of code," she conducted groundbreaking research at MIT Media Lab on racial and gender bias in AI and launched AJL to fight discrimination in algorithms. She frequently speaks at global forums (e.g. the UN, WEF) to promote equitable and accountable AI

Stuart J. Russell
https://www.linkedin.com/in/stuartjonathanrussell/
- Professor of Computer Science at University of California, Berkeley; holder of the Smith-Zadeh Chair in Engineering; Director of the Center for Human-Compatible AI
- Leading AI researcher focused on ethical AI alignment and safety. His work centers on the long-term future of AI and ensuring AI systems remain beneficial to humanity (He is co-author of the standard AI textbook Artificial Intelligence: A Modern Approach and an advisor on policy issues such as banning lethal autonomous weapons.)

Erik Brynjolfsson
https://www.linkedin.com/in/erikbrynjolfsson/
- Director of the Stanford Digital Economy Lab; Jerry Yang & Akiko Yamazaki Professor and Senior Fellow at Stanford Institute for Human-Centered AI (HAI), Senior Fellow at Stanford Institute for Economic Policy Research.
- Economist and author specializing in the impact of AI and digital technology on productivity, business, and the workforce. He has pioneered research on how IT and AI contribute to productivity and economic performance and co-authored the best-selling The Second Machine Age, examining work and prosperity in the AI era

Daron Acemoglu
https://economics.mit.edu/people/faculty/daron-acemoglu

- Institute Professor of Economics at MIT; Faculty Co-Director of MIT's Shaping the Future of Work Initiative
- Economist researching automation, AI, and their effects on labor, inequality, and economic growth. He advocates for redirecting AI development toward complementing workers rather than replacing them. Acemoglu's recent work (e.g. the book Power and Progress, co-authored with Simon Johnson) argues that AI and automation should be designed to broadly share benefits and improve worker productivity.

A Call to Reinvention

A Social Imperative

We have so little precedent for the now inevitable change we face. I've heard some compare it to the widespread adoption of the Internet in the late 1990s. At first, I thought so too, but now after my own reinvention, I think this misses something crucial. I lived, studied, and created my first software in the late 1990s, and this time feels entirely different. I've heard others compare AI to the advent of the smartphone and the explosion of mobile apps, or maybe the growth of Cloud services. I believe these examples are terribly inadequate too. Still others have gone further and compared it to the first use of electricity to drive early industry and innovation. This is closer, but it veils a deeper truth. To the best of my knowledge, electricity doesn't "think."

In a recent interview that hit my Youtube feed, the softspoken Geoffrey Hinton, who we've explored together, "hinted" that our best precedent is perhaps the dawn of the Industrial Revolution. He points out that this was when we last saw a human faculty replaced "outright" by machines. Man nor beast could compete with the machines being built to take on their physical tasks. Human "strength" was replaced forever by loud and often dangerous machines. Economic forces to drive efficiency and profit were at work then as they are now. The extraordinary investment in AI we're seeing today is, as we've said, to move it beyond its current, "narrow" stage, and to give it autonomy and agency. We are equipping an artificial being that possesses sound reasoning with the same tools that we, ourselves, would otherwise use to do our work. AI will use these tools, act on its decisions, and accomplish work that has historically required a human at the helm. And, it will do it better than most of us can. The implications of this are earthshattering. So what can the past teach us, if anything?

Well, just last week, I sat around a breakfast table with some very well-known and respected VCs and founders. These are good people who I've been able to interact with, from time-to-time, over the span of roughly 20 years. They're all brilliant, but equally down-to-earth and approachable. We'd had a flurry of emails the day before on the subject of Soaring Titan, which had prompted an otherwise impromptu gathering at the estate of one of these men. We sat there and did what business people do — got super excited talking about AI and all of its glorious business applications. Around that table, the topic of domestic manufacturing came up. There was a general consensus that the current Trump Administration's policies will likely drive up production in the "Good Old US of A." American factories and warehouses will be filled again — filled with beautifully buzzing autonomous machines. Not us. It might come as a surprise to many who've expected jobs to return for people, but it shouldn't be a shock. People are expensive. Very expensive. The same economic forces that drove jobs overseas to begin with will drive society towards more autonomous and capable machines, and all the better if these machines, and their patents, stay close to home. Some might argue — though none present at the meeting I attended — that we'll need people as technicians to maintain the machines, quality control managers, and so on. Don't count on it this time. The combination of fabricated intelligence and dexterity will very likely complete what was started 250 years ago. We are now seeing artificial limbs that look and operate very much like our own. This can revolutionize prosthetics, but at the same time, replace people on production lines that require careful handling of tools and materials. Why are such extraordinary investments being made? Curiosity? Kindness? Among the Seekers, perhaps; philanthropic Orchestrators, maybe. But extraordinary investments in enterprise are always made to realize multiples in return. Always.

Certainly those of us who have been around a while have seen automation grow in the workplace, and, sure, human replacement along the way has been a consequence for some. But most of us in our daily lives have not had to compete head-on with machines for our jobs; to preserve our livelihoods; or to feel self worth. For the most part, the pace of change has been slow enough that we've simply adapted and found more productive things to do. Likewise, the kind of automation that replaces human capital "outright" has been confined to a limited number of domains. But AI is crossing these domains. *All of them.* It levels the playing field because it's trained with data collected across these domains, "generally." To illustrate, let me describe another recent interview that caught my attention. I squinted my eyes, pursed my lips, and watched Sam Altman say that *ten years ago, most people would have expected "blue collar" jobs to be threatened first by AI, with "white collar" jobs following considerably later…* But, he continued — as if somehow surprised — that *now we see the reverse is true.* I found this amusing, if not a bit irritating, because Sam Altman knows very well why this is true. White collar workers have been creating the enormous datasets of intellectual property upon which his models are conveniently trained. So, of course the writers, software developers, and other creatives are the first to find themselves in a quandary. Don't get me wrong. I like Sam.

Evolving Social-Economic Structures

The integration of AI into the workforce is going to reshape how we organize society, and also how we think about human worth. It just will. It will do so in powerfully positive ways. And it will do so in powerfully destructive ways. Humans are, at their core, "explorers," and the human tradition is to explore both potentials to the full. What I write in the following paragraphs, I do as a "realist," a "centrist," an "independent voter," and a "free-thinker." I am not beholden to a given ideology or political "team." I really don't think that we have time for that kind of division, nor for those who gain by fostering it. We have big problems to solve — together. And now, the very concept of "together" must include both people who seek to advance their agenda and machines who seek to understand the people putting those agendas forward.

At the heart of today's transformation is a growing concern for structural unemployment, as job displacement might very-well outpace our ability to create new roles for us. Customer service representatives, logistics coordinators, and administrative professionals already find themselves outperformed by systems that can process, analyze, and respond with both superhuman efficiency and "human-enough" interpersonal skills, including convincing speech. Unlike previous technological revolutions, these changes aren't just shifting workers from one sector to another. They're eliminating entire categories of human labor. Again, I'm a career coder, but I'm also trusted to give business owners good advice, and it's quickly become difficult to defend the pricetags of traditional software development projects knowing what I know about AI's ability to code and to code well with the right grounding. Similarly, I'm finding myself teaching software development shops to use the technology for the first time because they've been too busy to notice how far it's come since they last dismissed it.

With job displacement, it's not hard to imagine an increasing divide in society. Picture an hourglass. At the top, you have the highly skilled AI specialists, researchers, and technologists who continue to command premium salaries. At the bottom, we see service workers in roles that still require human touch but see diminishing wages and security. In the all-important middle, where society's stability has always been maintained, we find a hollow core. This change won't affect all regions equally either. Communities built around manufacturing or service industries could face potential collapse, while tech hubs flourish. We could deepen divides that already tear at the fabric of society. Where the number of warehouses, data centers, and powerplants are likely to rise, we should expect to see them do so with increasing mechanisation.

At the same time, our traditional social safety nets, designed for a world of stable, long-term employment, are already showing signs of stress. Unemployment insurance, healthcare tied to full-time work, and pension systems assume a pattern of consistent employment that's rapidly becoming out-of-step with reality. In its place, we're seeing the rise of a kind of tech and AI-augmented "gig" economy; one where workers string together temporary engagements with little security and decreasing bargaining power. I'm not shy about saying that I've felt this personally. This shift fundamentally challenges the social contract between employers and employees that has historically underpinned successful economies. God knows that there needs to be a constructive dialogue about this between people and policy-makers with differing viewpoints — *bound together, first, by a sense of fraternity, sorority, and civic duty* — but that's not the system we have in 2025, is it?

Perhaps most concerning is the concentration of power in the hands of those who control AI systems and the data they require. Major tech companies, corporations, and people with vast resources are accumulating unprecedented influence. I like tech companies. I'm a technologist. I don't mind billionaires. I like ambition and systems that let people rise to their potential. But history teaches us to be pragmatic too. This concentration of power can create a "winner-takes-all" dynamic that stifles competition and innovation, killing opportunities for imaginative, ambitious, and well-meaning people like you and me. The power-shift also raises urgent questions about data privacy and civil liberties. Remember, our personal data is the fuel for these AI systems — systems that, themselves, will increasingly govern crucial decisions about our lives as we give them agency. AI is already making lending decisions, insurance decisions, and university admission decisions like those my kids now face.

On the topic of education, our systems remain largely rooted in industrial-era thinking. They were already struggling to adapt to 21st century challenges. The conventional model of front-loading education in our early years, by spending our youth in structured academic environments before transitioning to careers, is becoming misaligned with how the world really works these days. The skills that are likely to matter most in a world of intelligent machines are, as we've said, critical thinking, ethical judgment, and adaptability. Yet, these same skills often take a backseat to preparing for that next standardized test. Then there's access to a quality education to begin with. Who has it? What does it look like? Who pays for it? Inequality and poor decisions here threaten to create a permanent underclass of workers, completely unprepared for an AI-driven economy. Personally, I've been fortunate to provide a quality, private education to my children, but not because I believed that public education was anything less than the absolute best hope for addressing socioeconomic inequality in our time. Public or private, I think that both systems have aimed to turn Seekers into Makers with insufficient emphasis on becoming Orchestrators in life. I've watched my own children move from Seekers to Makers in the span of a few, short years — "pressing hard" to "achieve the grade." Yet, they will increasingly be faced with intelligent machines that can give a better answer every time. These machines are already operating at near-physicist level in some categories.

The psychological impact of these changes can't be understated, especially for our young — those enthusiastically heading off into a world that we should have prepared them for. Work has long been central to our human identity and social cohesion. The same is true for these kids. As automation reshapes or eliminates traditional roles, many face not just economic uncertainty but a crisis of personal purpose. I can tell you, and I know I'm not alone in saying it, there is nothing like the despair of wondering how you will provide for your family. It's a despair that fuels social unrest and populist movements, as people struggle to find meaning and agency.

The governance challenges are equally daunting. Our regulatory frameworks, designed for a slower-moving industrial system, are unlikely to keep pace with AI innovation. International cooperation would seem to be crucial and obvious, but that's quickly proving difficult as nations race to secure a competitive advantage in AI. Global inequalities could deepen as wealthy nations accelerate AI capabilities and developing regions fall further behind. As I write these words today, both the United States and the United Kingdom have refused to sign an international AI agreement at a global summit in Paris. I love and respect both of these nations. At the summit, they cited national security concerns, and because there will always be "bad actors," it's a position that's difficult to argue against. The agreement was signed by dozens of countries including France, China and India — all pledging an "open", "inclusive" and "ethical" approach to AI development. This is just what we need "in a perfect world," but whether any of the signers adhere to the agreement longterm is anyone's guess.

All of that said, AI-driven transformation does not have to be a story of gloom and doom. Let's assume that not all safeguards are lifted and AI is approached with a measure of common sense. Let's assume, further still, that people learn to approach one another again with an equal measure of common decency and purpose. Here, we find a great deal of hope.

New Potential Roles

We might see the emergence of AI Ethicists and other specialists who will help us ensure that machine decisions align with human values and ethical standards. These aren't just theoretical positions; they become a necessity in an AI-augmented business world, even if only to preserve the interests of shareholders. I say this because irresponsibility in business has always proven costly. New roles, like these, could help us build frameworks that balance human needs and harness machine intelligence concurrently.

Digital Ecosystem Architects could become the master builders of future workplaces, crafting environments where humans and AI systems work together in a harmonious coordination of effort. Think of them as the conductors of an orchestra, ensuring that every instrument — from natural language processing to robotics — plays its part in perfect synchronization. And who will teach us to play our own parts in this orchestra? Well, it's not hard to imagine human-AI collaboration coaches, serving as modern-day guides who help individuals and teams maximize their potential by working alongside intelligent machines.

The foundation of this AI-driven future rests on data, making roles like Data Trust Officers and Chief Data Curators likely to emerge. These guardians of information could help ensure that the lifeblood of AI systems remains pure, ethical, unbiased, and compliant with global standards, if and when we create them.

Meanwhile, AI Personalization Designers might emerge to craft experiences that feel uniquely human despite their algorithmic origins, whether in healthcare, education, or public services. As virtual reality and the metaverse evolve, we'll need integrators who can weave AI seamlessly into these virtual realms. They will create spaces that feel authentic and inclusive. These aren't just technical roles, they're creatives, architects, and artists of new social environments where human interaction can take on entirely new dimensions. Why not?

New Educational Systems

We might also see the emergence of more dynamic and responsive educational systems that recognize learning as a continuous, lifelong journey rather than a finite phase of life. This comes with the integration of AI systems that adapt and evolve alongside human learners. Imagine a near future where each person has access to an AI mentor that intimately understands their cognitive patterns, learning preferences, and developmental needs. These AI systems wouldn't just simply deliver standardized content and testing; they would craft personalized learning experiences that evolve in real-time based on each individual's progress, interests, and goals. For a student struggling with abstract mathematical concepts, their AI mentor might generate specialized visualizations or real-world examples that resonate with their particular interests, whether that's sports, music, or gaming. AI-driven educational systems could help bridge equity gaps in global education. In regions where access to skilled human teachers is limited, advanced AI mentors might provide high-quality, personalized instruction to students who would otherwise be left behind. Democratization of education through AI could catalyze a more equitable distribution of opportunities and resources that reasonable humans feel less threatened by. And, our AI agent friends would constantly evaluate modern conditions to keep these systems current and relevant for all of us. This evolution in education might prepare individuals not just for specific careers, but for a future where the ability to learn, unlearn, and relearn becomes the most valuable skill of all as we move from cradle to grave.

A Personal Mandate

What gives me the greatest hope, and what "moves me to emotion" as I watch AI being used by family and friends, is witnessing their individual discoveries — those moments when they realize that AI expanded the boundaries of what they thought was possible in their own creative exploration. To see people widen their creative horizon is exactly what moves an Orchestrator to flight.

So the personal mandate is simply this: never forget that before us, there was the thought of us; and that before anything we see made, there was the intention to make it. Anything that we manifest in our world must first take form in our own minds. These forms may be blurry at first, but eventually come into view with time, focus, and knowledge. Time, focus, and knowledge are personal limitations — limits exacerbated by an increasingly complex world. This complex world has robbed us of mystery and serendipity, with many left empty and unsatisfied without understanding why or to what extent. By pairing with machine intelligence, we may be able to cut a new path forward through these complexities, exceeding these limitations, and bringing the ideas that form in our minds to greater clarity — *faster*. Once these ideas begin to materialize in the physical world, machine effort can then further accelerate their transformation into the magnificent works that we — *as only humans can* — marvel at.

If we're going to create anything else worth marveling at in our time, then we must resist the temptation to outsource our imaginations to both "devices" and "divisive influences." Imagination is our superpower. It will sustain us in a world of intelligent machines if nourished. Albert Einstein once said that "imagination is more important than knowledge," though, in context, he prized knowledge as inspiration for his intuition. "Knowledge," he recognized, "is limited" — admitting that imagination is unlimited and therefore worth relying on even as (or, especially as) a theoretical physicist. I tell discouraged people that I've been blessed to observe that there does, in fact, exist "more light and beauty in the world than darkness and ugliness." But *imagination* is how that light and beauty was manifested by our forebearers and *imagination* is how they will be manifested for those who come after. That is a responsibility of the highest order.

To move from Maker to Orchestrator, we somehow must go back, relearn, and fully explore what it means to be a Seeker. How do we do this? Well, the answer isn't hidden from us, and yet we have such a hard time as adults finding it. The truth of it echoes through the ages, and we can read it plainly in our most sacred texts, while missing it entirely.

We must return to childlike wonder. We must seek moments and places that allow us to become deeply curious. We must, once again, let that childlike wonder and curiosity move us to explore diverse subjects; making it acceptable — *if not expected* — to be both scientist and philosopher. When we look up, we must be allowed to see the stars again. When we look down, we must take time to run our hands through the dirt again. When we're resting at home, we must observe the sunlight that shines through the trees, passes through our window, and dances on our wall like a performance made just for us again. We must return to walk the pilgrim's path again and again. We must see, kindly engage, and learn from our neighbors along the path we share again. We must reflect on the history beneath each footstep that we take again. And…

We must learn to "think" for ourselves again — *before fully embracing machines who "think" for themselves alongside us.*

In my darkest hour of need — *when it's 3am and my chest is collapsing in on itself* — if I'm disciplined to rise at that moment, "think" beyond my circumstances, and go to where my imagination lifts me, then I preserve my ability to "create." If I can create, then I can offer something of value — *to someone*. If I can offer something of value to someone, then I can "hope." *Thinking, creating, and hoping* are bound together. It's where we find meaning and purpose. It's also where we find meaning and purpose for AI in our world. AI formed in the murky, mysterious realm of the human mind — and not just one, but a collective. Its "lines and edges" gained clarity with time, focus, and knowledge. It then became real in our time. And so it must help each one of us, in kind, to do likewise — exploring and expanding our creativity, while rejecting our tendency to destroy and malign what is good, true, and beautiful.

Everything has its opposite and equal. If we're unable to gain altitude and to see beyond our circumstances, then we'll remain fixed in place and paralyzed by fear. This is true of an individual and it is true of a society. Let's hope it does not become true for intelligent machines. When we remain fixed in place and paralyzed by fear, our thoughts, nonetheless, press on ahead of us. Alone — *stuck, frozen, and angry* — we'll contemplate doom. If we think of it long enough, then we will create it, and once created, we're hopeless. We're hopeless because of its emptiness — *being of nothing material* — and offering no material value as a result. It will remain an apparition, relentlessly haunting us as the specter of what might have otherwise materialized by Faith.

Philosophy of the Soaring Titan

For two years straight, I committed myself to a daily recitation. I authored and memorized its six paragraphs. The prose grew in length and depth over time. It was my manifesto — my crutch. I'd tape each revision to the wall beside my desk and read it by candlelight before the sun rose. Through its words, I began to reclaim what it meant to be a Seeker and to press on towards an Orchestrator. It was called, "My Definite Major Purpose." Carol didn't like this title, but I couldn't claim credit for it anyway. I'd lifted the title from one of two classical business texts that were given to me by a much older friend and mentor in 2012. His name was Dr. Ralph Coffman, and the texts were written by Napoleon Hill.

At the time, Ralph told me that he didn't often share these books with people. Inside the cover of "Think and Grow Rich," he wrote that he believed I would read *and re-read* these books until I discovered their secrets. Ralph had an incredible story... He'd been a Golden Gloves boxing champion, a doctor, a member of President Lindon Johnson's administration, worked for Governor Wendell Ford in Kentucky, founded and exited a very successful healthcare concern, and served — *humbly* — in the church my family was then attending. For some bizarre reason that I still can't understand, I'd been asked to be the Finance Chairman at that church during a very difficult time in its financial history. Ralph was kind, reserved, and helpful to a then 35 year old with a young family, and a great deal yet to learn.

The six paragraphs that I wrote were inspired by what I'd read and re-read in those books. As part of my reinvention in 2023, I was determined to figure out just what Ralph saw in those books — and in me. In my daily recitation, I would state what I wanted as my "definite major purpose;" follow it by my "working plan;" describe how I was going to "go the extra mile" to achieve it; remind myself of who belonged to my "mastermind team" and why; revisit the "habits and self-discipline" I would automate; and recommit myself to "applied Faith."

At some point during those two years — as I contemplated a new venture with Michael Browning called Soaring Titan — I realized that what had become nearly five pages of text could be condensed into this single philosophy.

> *"The Soaring Titan knows what he wants, why he wants it, and how to orchestrate its achievement though a symbiotic collective of minds (and machines) – bound together in a harmonious coordination of effort. He is mortal, but he embraces and exercises the immortal Faith necessary to live each day as though he's already achieved his vision. This is his superpower – leaving no alternative for his mind, or the universe, to accept as reality – automating the required patterns that sustain his success habitually. The Soaring Titan knows who he is, what he's made of, where he came from, and to where he is ultimately returning. He rejects fear and chooses to Think his wings into being by Faith – wings that will not melt as he crosses the sky."*

What does it mean? This might sound cliché, but you'll have to figure it out for yourself. It might take years or it might come quickly. Your expression of it will likely differ and become your own, but the truth of it is ancient and immutable. Here is a hint — taken from a blog post that I've hesitated to post. It's been too personal, covering my 20 years of seeking, making, and orchestrating life as a husband and a father ("Adventure of a Lifetime at the Speed of Sound").

We're born believing in magic. We enter life with a hunger for storytelling and the best stories never lack it. The impossible becomes possible! The underdog wins! Good triumphs over evil!

Points come in your life when you must choose the story that you want to tell when you become the storyteller. That's what life is! If you project the story forward, and imagine every day "THIS is the story I'm telling in the future," then you'll make the plans, commitments, adjustments, and connections required to walk the distance that becomes the story. And, you will receive help along the way that is undeniably… magical. **This is the Secret of the Master Key.**

And so when I look back on those moments that were truly magical — *moments that should never have been possible* — I realize, now, that I had already pictured them in my mind (quite literally) every day well-beforehand. I would play them in my head over and over again without even realizing what I was doing. They would keep me company on my long walks. I would score them with music as they took shape, causing each scene to lift me to high emotion. I knew exactly how each scene would play out. I knew exactly how I would present the collection of scenes for others and the effect it would have. There was no question of success or failure. These moments I pictured and planned were always about the family… About my Carol. About my Gabriel. My Jude. And my Sarah Kate… About us living, loving, and learning together somewhere in the world… Experiencing the Good, the True, and the Beautiful together. But as the kids now leave to pursue their own lives, I am reminded that they are not mine and never were. The memories, however, are mine. The story is and forever will be a part of me — in this life and beyond.

What will your story be?

There is truth and power in it.

Additional Reading

AI DEEP RESEARCH SUGGESTIONS

Title	Author	Year
Leibniz: An Intellectual Biography	Maria Rosa Antognazza	2009
Leibniz (The Routledge Philosophers)	Nicholas Jolley	2005
Gottfried Wilhelm Leibniz: The Polymath Who Brought Us Calculus	M. B. W. Tent	2012
The Best of All Possible Worlds: A Life of Leibniz in Seven Pivotal Days	Michael Kempe (trans. Marshall Yarbrough)	2022/2024
Alan Turing: The Enigma	Andrew Hodges	1983
Alan Turing: Life and Legacy of a Great Thinker	Christof Teuscher (ed.)	2004
Turing: Pioneer of the Information Age	B. Jack Copeland	2012
Alan Turing: His Work and Impact	S. Barry Cooper & Jan van Leeuwen (eds.)	2013
A Mind At The Forefront of AI: The Fascinating Life of Geoffrey Hinton, His Unexpected Departure From Google, His Predictions And Warnings.	Steven Wright	2023
Genius Makers: The Mavericks Who Brought AI to Google, Facebook, and the World	Cade Metz	2021
Deep Learning	Yann LeCun, Yoshua Bengio, Geoffrey Hinton	2015

Index

AI Agency	3, 7, 68, 77, 111, 125
AI Ethicists	129
AI Ethics	123
AI Ethics Auditor	115
AI Explainability	115
AI Fairness	123
AI Models	42, 86
AI Orchestrator	118
AI Personalization Designers	129
AI Workflow Orchestration	92
AI accessibility tool	95
AI agent	109, 110, 112, 121
AI agents	32
AI automation	30
AI-Assisted Imagining	35
AI-Assisted Reflection System	92
AI-Based Resilience Coach	92
AI-driven educational systems	129
AI-powered monitor	94
AI21	17
AWS Simple Email Service	97
Adaptability	128
Adobe	18
Affectiva	98
AgentGPT	20
Alan Turing	6, 24, 68, 91, 96
Albert Einstein	130
AlexNet	13
Alexa	15
Algorithmic Bias	123
Alicante	1

Alina	1
Alonzo Church	6
Amazon	18
Amazon Polly	18
Amsterdam	1
Ancient Greece	4
Andrew Hodges	134
Andrew Ng	65
Anna Sud	69
Anthropic	18, 79, 81, 84, 89, 96, 120, 126
Anthropic AI	80
Anthropic Claude 3.5	17
Arthur Samuel	9
Artificial General Intelligence	23
Artificial Intelligence	1, 5, 6, 7, 8, 11, 15, 21, 25, 36, 51, 90, 100, 123
Artificial Superintelligence	25
Asana	22
Assistant Roles	77
Attention Mechanism	41
AutoGPT	37
AutoGen	20
Automation	128
Autonomous Agents	22
Autonomous Machines	125
B. Jack Copeland	134
BERT	14
Banks	15
Bard	19
BaseAgent	117
BaseTool	117
BenchLLM	85
Bias Bouncer	38
Big Data	12
Binary System	36
Biomechatronics	98
Bletchley Park	6, 32, 69, 88, 96

Bloch Sphere	54
Blue Origin	103
Bombe machine	66, 67
Business Experimentation Agent	122
CRM	21
Cade Metz	134
Cambridge	6
Cambridge University Press	52
Camden	1
Carnegie Mellon	7
Carol Turner	1, 37, 63, 64, 131, 133
Chain of Thought	44
ChatGPT	19, 29, 45, 55, 80, 88, 102
Checkers Player	9
Chief Data Curators	129
China	128
Christof Teuscher	134
Christopher Reeve	33
Church-Turing Thesis	6
Clark Joseph Kent	33
Clark Kent	33
Classical Computing	50
Claude	29, 45, 57, 62, 82
Claude Shannon	8
Claude Sonnet	61, 74, 120
Claude Sonnet 3.5	61
Claude Sonnet 3.7	120
Cohere	24
Collective Intelligence	99
Communication System	112
Compestella, Spain	1
Computation and Logic	66
Computer Science	4
Confluence	22
Consumer-facing LLM	45

Context Window	42
Convolutional Neural Networks	13
Crete	4
CrewAI	20
Critical Thinking	128
Cryptanalysis	69
Cybersecurity	69
Cynthia Breazeal	98
DARPA	10
DENDRAL	9
Daron Acemoglu	123
Dartmouth College	7
Dartmouth Conference	10
Data Science	4
Data loss scenarios	71
David Eagleman	98
Deep Learning	15, 39, 101, 114
Deep Learning Boom	15
Deep Research	44, 123
DeepSeek	21
Deutsch-Jozsa	49
Diagrams	87
Diagrams.net	86
Digital Ecosystem Architects	129
Digital twins	59
Distillation	40
Docker	88
Dr. Ralph Coffman	132
Dragon NaturallySpeaking	12
Dunning-Kruger Effect	2, 62
Dynamic Task Prioritization	112, 117
E. Rieffel	47, 51
ELIZA	9
ERP	26
EU	6

Economic Growth	124
Edison	11
Eleanor Rieffel	47
ElevenLabs	18
Elon Musk	102
Embeddings	41
Emergency "stops"	113
Emotion AI	98
Empatica	98
Encryption Protocols	69
Enigma	69
Enigma code	69
Entrepreneurial Agent	122
Eric Johnston	47
Erik Brynjolfsson	123
Error detection and reporting mechanisms	113
Ethical AI Applications	98, 115
Ethical Judgment	128
Europe	6, 21
Evals	85
Evolving Social-Economic Structures	126
Execution Engine	118
Explainable AI	115
Fei-Fei Li	65
Few-shot Learning	43, 72
Fine-tuning	40
Fortress of Solitude	33
France	128
Function Calling	76, 109
GPT	18
GPT-4o	82, 83
Gabriel Ryan Turner	38
Garry Kasparov	12
Gemini	29, 45, 55, 79
General Motors	70

Generative AI	42, 70, 84
Geoff Mulgan	99
Geoffrey Hinton	101, 125
Gig economy	127
GitGuardian	85
GitHub	17
GitHub Copilot	28, 84
Gmail API	97
Google	18, 82, 103, 120, 123
Google Cirq	47
Google Cloud TTS	18
Google Gemini	18
Google Sheets	97
Google Speech-to-Text	95
Gottfried	36
Gottfried Leibniz	36
Gottfried Wilhelm Leibniz	5, 36
Graphics Processing Units	13
Grover's algorithm	49
Haiku	74
Hallucination in AI	42
Hanover	7
HashiCorp Vault	85
Healthcare Systems	22, 129
Hinton	62, 116, 125
Hugging Face	85
HuggingFace Multimodal pipelines	19
Hugh Herr	98
Human-AI Collaboration	116
Human-AI Collaboration Coaches	129
Human-Centered AI	65
Hut 8	67
I. L. Chuang	51
IBM	53
IBM Quantum	51, 53

IBM Quantum Composer	53
IBM Quantum Experience	51, 53
IBM's Deep Blue	12
Imitation Game	6
India	128
Industrial Revolution	125
Inference	40
Information Age	134
Input Sanitization	82
Intelligent Machines	2
Isaac Chuang	47, 48, 51
Ivan	2
JSON	92, 109
JSON Schema	74
JSON Storage	92
JSON format	71
James Cameron	45
James Lighthill	10
Jan van Leeuwen	134
Jira	22
Joan Clarke	68
John McCarthy	8
John von Neumann	6
Jor-El	33
Joseph Weizenbaum	9
Joy Buolamwini	123
Jude Nicolas Turner	1, 63, 64, 133
Kal-El	33
Kentish Town	1
Kentucky	33
Kiev University	69
LLM	50, 63, 68, 97
Labor	129
LangChain	20, 85
LangSmith	85
Langflow	20

Large Language Model	105
Large Language Model (LLM)	105
Large Language Models	25, 32, 36, 96
Le Cordon Bleu, London	1
Learning Journey	49
Leibniz	37, 52, 61, 62
Leipzig	36
Lighthill Report	10
Lindon Johnson	132
Lisbon	1
Llama	76, 77
Llama 3	77
LlamaIndex	17, 84
Local Development Environment	88
London	1, 37
Louisville	104
Louisville, Kentucky	103
M. A. Nielsen	51
M. B. W. Tent	134
MIT	11, 41, 42, 53, 128
MIT Media Lab	98, 123
MIT OpenCourseWare	47
MIT Press	51
MIT Sloan School of Management	99
MYCIN	11
Mac Hack	9
Machine Intelligence	56, 130
Machine Learning	13, 39, 51
Maria Gimeno-Segovia	47
Maria Rosa Antognazza	134
Marshall Yarbrough	134
Marvin Minsky	8
Maslow's Hierarchy of Needs	30

Massachusetts Institute of Technology (MIT)	38
Mastering AI Agency	64, 116
Memory and Learning Architecture	112
Meta	18
Meta ImageBind	18
Michael Browning	1
Michael Kempe	134
Michael Nielsen	47, 48, 51
Microsoft	2, 102
Microsoft Azure TTS	18
Microsoft Kosmos-1	18
Microsoft Presidio	86
Microsoft Research	20
Microsoft Teams	22
Mingrammer	86
Mistral	76
MkDocs	86
MockGPT	85
Moesif	86
Moore's Law	12
Muhammad Ali	104
Murf.ai	18
Mustafa Suleyman	65
N. D. Mermin	52
NVIDIA	18
Napoleon Hill	132
Nathan Rochester	8
Natural Language Processing	10, 16
Netflix	16
Neural Architecture	106
Neural Networks	39, 102, 116
Neuralink	29
New Hampshire	7
New Shepard spacecraft	103

Nic Harrigan	47
Nicholas Jolley	134
Nielsen & Chuang	48
Normandy	1
Odessa	69
Onovative	2
OpenAI	21, 35, 64, 74, 79, 81, 83, 84, 85, 88, 89, 91, 93, 95, 96, 98, 102, 109, 113, 120, 123
OpenAI GPT-4V	18
OpenAI Usage API	85
OpenAI's Whisper	95
OpenAPI/Swagger	86
OpenCV	94
Operational boundaries	113
Opus	74
Orchestrator	105
Orville Wright	34
PROSPECTOR	11
Paris	128
Pattie Maes	65
Peter Shor	46
Pivot & Adapt Framework	121
Play.ht	18
Porto	1
Predictive analytics	15
Prompt Engineering	42
PyCharm	84
PyTorch	14
Python	37, 91, 120
Python Environment	96
Python programming tutor	71
Qiskit	53
Qiskit textbook	53
Quantum Bits Qubits	46
Quantum Country	53
Quantum Fourier Transform	49

Quantum Programming Frameworks	49
Quantum error correction	49
RAG applications	78
RPA	26
Rana el Kaliouby	98
Reasoning Models	44
Recurrent Neural Networks	13
René Descartes	4
Replit	18
Resemble AI	18
Retrieval Augmented Generation (RAG)	42, 78
Richard Greenblatt	9
Robert Oppenheimer	100
Rosalind Picard	98
Rose Luckin	65
Runway ML	18
S. Barry Cooper	134
SaaS	2
Safety and Control Mechanisms	112
Sagres, Portugal	1
Salesforce	2
Sam Altman	126
Sarah Kate Turner	1, 61, 63, 64, 133
Semantic Search	78
SendGrid	97
Seville, Spain	1
Shaping the Future of Work Initiative	124
SharePoint	22
Shor's algorithm	46
Simon Johnson	124
Sir Isaac Newton	36
Siri	15
Slack	22

Smart city platforms	59
Soaring Titan	2, 93, 114
Social Robotics	98
Sphinx	86
Spiking Neural Networks	25
Spock	105
Spock's Brain	105
Stanford	7
Stanford Digital Economy Lab	123
Stanford University	37, 65
Star Trek	105
Stepped Reckoner	45
Steve Wozniak	102
Steven Wright	134
Stuart J. Russell	123
Sundar Pichai	102
Superhuman	7, 63, 66, 67, 68, 77, 79, 80, 95, 110, 115, 126
Superhuman Formation	35
Superintelligent AI	27
Superminds	99
Support Vector Machines	13
Surface Codes	47
Symbolic AI	8
Synchron	29
System Instructions	73
Systems thinking	60
Talos	5
Technological disruption	60
Technology	1, 2, 8, 11, 13, 14, 15, 16, 19, 21, 29, 32, 33, 34, 37, 51, 59, 60, 61, 62, 64, 98, 101, 102, 105, 123, 127
Temperature	75
Temperature and Sampling	42
TensorFlow	14
Terminator	45
Terrell	103
Text-to-Video	18

TextBlob	92
The Bloch Sphere	54
The New York Times	103
Think and Grow Rich	132
Thomas Hobbes	5
Thomas W. Malone	99
Tiktoken	85
Timnit Gebru	123
Tokens	42
Tool Use	75
Top-P	75
Training Data	39
Transformer	13
Transformer Architecture	41
Transparent Decision-Making	43
Transportation and Mobility	57
Turing	66
Turing Award	101
Turing Test	6, 68
Turing machine	6
UCI Machine Learning Repository	12
US	6
Ukraine	69
United Kingdom	1, 128
United States	128
University College London	99
University College London (UCL)	65
University of California, Berkeley	123
University of Louisville	69
User Roles	73
V7 Labs	85
VADER	92
Validation checks	113

Vector Databases	84
Vector Embeddings	92
Vectorization	41
Vision API	94
Visual Studio Code	84
W. Polak	51
WellSaid Labs	18
Wendell Ford	132
Wikipedia	61
William Shatner	103
Winston Churchill	67
WireMock	85
Wright brothers	11
YAML	86
YOLO	94
Yann LeCun	134
Yoshua Bengio	134
YouTube	16, 61
Zero-shot Learning	43
computer vision	13
data privacy	127
digital tide	64
dynamic pricing	58
embedding techniques	78
exercise_9.py	117
goal framework	112, 118
minutephysics	54
multi-modal cognition	24
o1 Pro	120
personalized instruction	129
predictive maintenance	58
pydantic	120
quantum computing	50, 52
reasoning model	62
smart building system	108
smart home system	106

smart sensors	58
smartphone	64
superhuman powers	33
symbolic logic	7
tool integration	108
traffic management	58

www.ingramcontent.com/pod-product-compliance
Lightning Source LLC
Chambersburg PA
CBHW080442110426
42743CB00016B/3252